D0019842

Spark Your Child's Success

in Math and Science

Practical Advice for Parents

JACQUELINE BARBER,
NICOLE PARIZEAU, LINCOLN BERGMAN

ILLUSTRATIONS BY PATRICIA LIMA

LAWRENCE HALL OF SCIENCE
UNIVERSITY OF CALIFORNIA AT BERKELEY

GREAT EXPLORATIONS IN MATH AND SCIENCE (GEMS) • BERKELEY

Lawrence Hall of Science

LHS GEMS®

Great Explorations in Math and Science (GEMS)
University of California at Berkeley
Berkeley, CA 94720

Distributed to the trade in the United States and Canada
by Consortium Book Sales & Distribution.

Design by Stefan Gutermuth/double-u-gee

Library of Congress Cataloging-in-Publication Data

Barber, Jacqueline.
  Spark your child's success in math and science : practical advice for
parents / Jacqueline Barber, Nicole Parizeau, Lincoln Bergman;
illustrations by Patricia Lima.
     p. cm.
Includes bibliographical references and index.
  ISBN 0-924886-71-4 (alk. paper)
  1. Mathematics--Study and teaching--Parent participation. 2.
Science--Study and teaching--Parent participation. I. Parizeau, Nicole,
1957- II. Bergman, Lincoln. III. GEMS (Project) IV. Title.
  QA11.2 .B35 2002
  649'.5--dc21
                                    2002002433

JB—For my mother, Emma Lou Orth, whose own infectious sense of inquiry and adventure helped shape me, and in turn my three children, Saul, Leib, and Jesse.

NP—For Monique Parizeau, my mother, whose profound curiosity about the world galvanized my childhood; for my second families, the Van Vurens and the Gaugs; and for Jeff.

LB—For my parents, Anne and Leibel; for the two wondrous daughters of my heart, Anna and Caitlin; and for all the world's children—the sunrise in the morning of a better day.

Acknowledgments

We are indebted to many educational and editorial colleagues for making this book possible. Thanks, first, to the tireless educators and researchers at the Lawrence Hall of Science, whose decades of experience and innovation were the source of so many suggestions in this book. We also thank Lynn Barakos and all who worked on the GEMS handbook *Parent Partners: Workshops to Foster School/Home/Family Partnerships*, which launched our efforts to involve activist parents in their children's schooling and served as a springboard to this book. *A Parent's Guide to Great Explorations in Math and Science* is a quaint (and out-of-print) resource now, a legacy of our publishing infancy, but it was the germ of an idea that matured over the years and culminated in this book; for that inspiration, thanks to all who participated. Several sterling educators contributed anecdotes to these pages, illustrating a number of our ideas with personal gems; many thanks to Katharine Barrett, John Erickson, Barbara Nagle, and Carolyn Willard.

Art director Stefan Gutermuth was an endless source of professionalism and patience during the creation of *Spark*, and lent his considerable talent to both the original vision for this book and its design. Stefan, it was a pleasure. Florence Stone brought her eagle eye to the task of proofing the book, provided much-appreciated suggestions on its consistency, and tackled the resources and index with dedication. We thank Steve Dunphy, Matthew Osborn, and Deborah Lee Rose, who brought fervor to *Spark*'s promotion and radiated faith in its success. We greatly appreciate the time and suggestions of those who read earlier drafts of the book: Marian Diamond, Karen Ostlund, Monique Sicotte, Lisa Simpson, Michael K. Stone, Steve Sutcher, and Jeffery Wilcox.

The authors are particularly proud and fortunate to work with the spirited team at Great Explorations in Math and Science. GEMS is a collective of talent and soul; for the integrity and support of everyone in the team, thank you.

This book's development was supported in part by a generous grant from the William K. Holt Foundation.

From *Uncle Tungsten: Memories of a Chemical Boyhood*

BY OLIVER SACKS, DISTINGUISHED NEUROLOGIST AND AUTHOR

I loved light, especially the lighting of the shabbas candles on Friday nights, when my mother would murmur a prayer as she lit them. I was not allowed to touch them once they were lit—they were sacred, I was told, their flames were holy, not to be fiddled with. I was mesmerized by the little cone of blue flame at the candle's center—why was it blue? Our house had coal fires, and I would often gaze into the heart of a fire, watching it go from a dim red glow to orange, to yellow, and then I would blow on it with the bellows until it glowed almost white-hot. If it got hot enough, I wondered, would it blaze blue, be blue-hot?

Did the sun and stars burn in the same way? Why did they never go out? What were they made of? I was reassured when I learned that the core of the earth consisted of a great ball of iron—this sounded solid, something one could depend on. And I was pleased when I was told that we ourselves were made of the very same elements as composed the sun and stars, that some of my atoms might once have been in a distant star. But it frightened me too, made me feel that my atoms were only on loan and might fly apart at any time, fly away like the fine talcum powder I saw in the bathroom.

I badgered my parents constantly with questions. Where did color come from?

Why did my mother use the platinum loop that hung above the stove to cause the gas burner to catch fire? What happened to the sugar when one stirred it into the tea? Where did it go? Why did water bubble when it boiled? (I liked to watch water set to boil on the stove, to see it quivering with heat before it burst into bubbles.)

My mother showed me other wonders. She had a necklace of polished yellow pieces of amber, and she showed me how, when she rubbed them, tiny pieces of paper would fly up and stick to them. Or she would put the electrified amber against my ear, and I would hear and feel a tiny snap, a spark.

...My questions were endless, and touched on everything, though they tended to circle around, again and again, to my obsession, the metals. Why were they shiny? Why smooth? Why cool? Why hard? Why heavy? Why did they bend, not break? Why did they ring? Why could two soft metals like zinc and copper, or tin and copper, combine to produce a harder metal? What gave gold its goldness, and why did it never tarnish? My mother was patient, for the most part, and tried to explain, but eventually, when I exhausted her patience, she would say, "That's all I can tell you—you'll have to quiz Uncle Dave to learn more."

CONTENTS

RESOURCE BOXES

Spark Your
Child's Success
in Math and Science

Introduction

Why This Book?

**A parent's involvement in a child's education is the single
most important factor in that child's academic success.**

The single most important factor. Decades of educational research tell us that
an involved parent contributes overwhelmingly to her child's grades and test
scores, school attendance and quality of homework, positive attitudes and
behavior at school, likelihood of graduation, and desire to enroll in higher
education. In many ways, as we'll describe, you're the essence of your child's
education; you've got the power!

It's true that many other important factors, including school funding,
teacher qualification, student resources, child nutrition, and a host of others
can swamp our considerations of what affects academic success. These factors
do matter. A lot. But research about the family's role, a parent's influence,
and the relationship between school and home has produced clear-as-a-bell
results: **nothing affects the academic outcome for a child as much as the
involvement of a parent or other adult caregiver in that child's education.** This is true no matter what personal factors are at work: the number of
parents raising a child; the family's economic situation; the parent's familiar-
ity with English; the size of the family; the parent's education; or a child's
own interests, talents, and abilities. The bottom line is that whatever your
academic or cultural background, your family situation, or the many pulls on
your time, you are in the most influential position to shape your child's
future.

Parents and other adult caregivers are their children's first and most endur-
ing teachers. Even the best teacher your child encounters in school will only
be with your child for a year, or perhaps two; even after children enter school,
they spend seventy percent of their waking hours outside of the school setting. As a parent, you have greater opportunity to make a difference, to
teach, model, and guide your child's learning, than anyone else. You have a
more intimate knowledge of your child's needs and talents. You have a keen-
er interest in your child's schooling and future, and deeper motivation to help
your child succeed. No one is better placed or more qualified than you to
make a difference in your child's academic and lifelong education.

Parent involvement is also strongly associated with effective schools.
Teacher morale soars, the school's community reputation goes up, and the

A family can take many legitimate forms. In this book we use the word "parent" to mean any adult caregiver who plays a parenting role for a child. That may be a biological parent or a stepparent, aunt, grandparent, or unrelated adult living in a child's household.

school's academic standings rise demonstrably when parents get involved with their children's schooling. It can be a humbling experience to realize the academic and social influence you have on your child and her school. It can feel…BIG. We've written this book to help make the entire experience of interacting with your child, her teachers, and her schools productive, UNintimidating, and even exciting. To start with, the feelings that brought you to this book in the first place indicate that you're already an engaged and involved parent. We hope to share clear, positive suggestions to build on what you already do as a caring and involved parent, caregiver, or adult role model. In that spirit, we welcome you to these pages.

What's in This Book?

This book offers a variety of ways in which to deepen your involvement with your child's education. We hope you'll read it in whatever way makes the most sense to *you:* as a whole, to understand the overarching nature and direction of education today and your immensely important role in it, or piecemeal, according to what suits your needs and inclinations. The chapters are broken down into small, manageable ideas to make it easy to spot topics of special interest. Important ideas are sometimes repeated, to give you every chance to catch and absorb them. And you'll find lots of special notes and highlighted information for practical use.

In Chapter One, **A Parent's Role: The Most Important Factor of All**, you'll read about the differences between schooling and education, about the importance of your connection to your child's school and teachers, and about the very meaning of academic "success" for your child. You'll see how, as parents, we sometimes inadvertently send mixed messages about schooling to our children. We'll describe parent involvement clearly—what it takes, what it *doesn't* require of you, and how to begin where you feel most comfortable.

Chapter Two, **Math and Science Education Today: What's New Since We Were in School?** explores the galaxy of current information about education. We'll share how research is guiding educational reform in some exciting new directions as studies reveal more about how children learn best. We'll describe that research, and explain what's changed in educational thinking since our own school days. You'll learn about best teaching and assessment practices, and about testing. We'll go behind all the talk about "standards" and clarify how they affect what our children learn. We'll define and demystify educational jargon so you can concentrate on the *content* of changes going on in education, not the vocabulary.

In Chapter Three, **Proven Strategies: Supporting Your Child's Success in Math and Science**, we'll do our best to help you decide what's truly important in the spectrum of education today as it relates to *you*. You'll be introduced to the idea that a child has many kinds of intelligence (as we all do), and that her learning style can affect her entire education. Parenting styles differ too; we'll help you identify your style and what it may mean for how your child learns. We'll explain how the relationships you form with your child's school and teachers have a huge impact not only on your own child's success, but on the success of all the students in the school. The issue of gender, as you can imagine, is very important for today's child, and we'll describe the most current research findings on whether and how gender may affect learning. In this chapter we also note that no matter how "wired" the country appears to be these days, access to technology is far from universal; we'll discuss the importance of that gap and alternatives to having a computer at home. And yes, we'll talk about homework—whether and how much it matters, and how to cope.

Chapter Four, **The Spirit of Inquiry: Keeping the Scientist Alive in Your Child**, gets to the heart of **inquiry-based learning**—the most significant idea in mathematics and science education today. Putting inquiry into practice can positively affect not only your child's success in school, but how he approaches learning and education for the rest of his days. We'll describe inquiry-based learning and show you how to apply its principles right away, anywhere, to support your child's success in math and science, and explain how valuable this absorbing and effective approach can be. A good first description is that inquiry-based learning cultivates and capitalizes on a child's natural curiosity to help her learn about the world. It's about encouraging your child's curiosity and natural wonder, and fostering a fascination for discovery and a way of learning that has real-world relevance. The slogan "every child a scientist" is an exponent of inquiry-based learning, which is both child-centered and based on the real ways in which scientists learn about the world.

The *National Science Education Standards,* the *Principles and Standards for School Mathematics,* and the *Benchmarks for Science Literacy* advocate precisely the kinds of programs the Lawrence Hall of Science develops. LHS not only addresses standards your children's schools are required to meet, but bases its materials and programs on current educational criteria and research—and makes them useful and clear for all educators of school-aged children.

Where We're Coming From

The University of California at Berkeley's Lawrence Hall of Science (LHS) perches high in the Berkeley Hills overlooking the San Francisco Bay Area. The sweeping view could be a metaphor for the hall itself; LHS is:

- **a public science center**, offering a full program of family workshops, after-school and summer classes, exhibits, films, lectures, and special events;

- **a center for school change**, providing ongoing professional development opportunities for teachers of preschool through twelfth grade, and building long-term relationships with schools to improve curriculum and instruction, increase parent involvement, and help create a reflective culture within schools;

- **a center for curriculum innovation**, engaging in research and development of instructional materials in mathematics and science for students, teachers, administrators, parents, and families. Instructional materials developed at LHS are in use in all fifty states and in over twenty percent of the country's elementary schools, and have been translated into twenty-seven languages to date.

The educational contributions of the Lawrence Hall of Science make it one of the most highly regarded public science centers in the country. LHS has a long-standing reputation for developing and promoting inquiry-based learning, an approach that centers on the learner and her own curiosity. This approach infuses all the curriculum materials LHS creates and the programs it offers. Since the 1960s, along with other leaders in the field, LHS has helped position **inquiry-based learning** at the heart of most reform efforts in the field of science education. And LHS is right in the thick of things when it comes to national and state academic standards, working with schools and teachers to help students reach or exceed these with the use of inquiry-based instruction. (See page 56 and Appendix A for more information on standards.)

The Lawrence Hall of Science EQUALS program was a pioneer in understanding that families are critical to the academic success of children. In the early 1980s, EQUALS made a great breakthrough in making mathematics more accessible to all children through its FAMILY MATH program. The work of FAMILY MATH is built on the need to bridge the gap between the cultures of home and school. With the creation of FAMILY MATH, the floodgates opened and the family-programs movement was ignited nationwide. The Lawrence Hall of Science helped launch FAMILY SCIENCE in the late 1980s, developed an active FAMILY HEALTH program in the '90s, and has created an array of other family and parent programming and materials. These include:

> *National Science Education Standards*, page 23:
> Inquiry is a multifaceted activity that involves making observations; posing questions; examining books and other sources of information to see what is already known; planning investigations; reviewing what is already known in light of experimental evidence; using tools to gather, analyze, and interpret data; proposing answers, explanations, and predictions; and communicating the results. Inquiry requires identification of assumptions, use of critical and logical thinking, and consideration of alternative explanations.

- **Parent Partners**, a parent-education program designed to involve parents in their children's education
- **Guides** designed to help parents lead their children in learning about the natural world
- **Parent Portal (www.lhsparent.org)**, a Web site devoted to practical tips and tools for parents of school-aged children
- **Home Science and Mathematics Activity Kits**, commercially available through the Scientific Explorer product line

These and other family/parent materials from LHS have been extraordinarily successful. FAMILY MATH, the grandmother of family and parent programs at LHS, is now available in four languages and used by over five million families around the world. For some examples of Lawrence Hall of Science offerings developed for (or readily adaptable by) parents of school-aged children, as well as a couple of like-minded programs we admire, please see Appendix B on page 130.

The determination to make its programs and materials accessible to everyone makes the Lawrence Hall of Science a natural partner for parents. We work hard to understand learning, and we're especially aware of the fertile ground every family can provide for a youngster's growth.

You Are the Key!

With this book, we offer you clear, jargon-free, and practical advice on ways to play the most important educational role in your child's life. From the lessons we've taught (and learned, and are always learning!), the research we've digested, the examples we've created, and the scope of our work, the Lawrence Hall of Science is perfectly placed to help parents make sense of and contribute to their children's success in math and science, and to sustain their educational achievements overall.

As a parent, you are the best person in the world to spark, sustain, or renew your child's sense of discovery and excitement during her school years. And **whatever** your background or experience, you're the best person to support your child in math and science—at school and outside of school. In these pages we'll show you how to navigate educational issues; understand your crucial role in your child's academic success; and put inquiry-based learning into practice with your child, to light and sustain the flame for learning.

MATH AND SCIENCE

CHAPTER ONE

A PARENT'S ROLE
The Most Important Factor of All

What Do We Mean by "Parent Involvement"?

What the research says about parent involvement

Research about the role of parents and the family environment in student achievement is fueling a transformation in many schools. The editors of *A New Generation of Evidence* (Anne T. Henderson and Nancy Berla, eds.; Center for Law and Education, 1994) call this blossoming research "a growth industry." There are books about it on shelves everywhere, and teachers, administrators, and parents themselves are much more aware of the interconnectedness of home and school than in the past. **As a parent, your involvement is of greater benefit to your child's academic success than was ever imagined**. This isn't just a trend or hypothesis; overwhelming research evidence bears this out.

When schools work together with families to support learning, children tend to succeed not just in school, but throughout life.

RESEARCH SHOWS THAT THE FOLLOWING THINGS, NOT INCOME OR SOCIAL STATUS, ARE THE BEST PREDICTORS OF A STUDENT'S ACHIEVEMENT IN SCHOOL

1. A home environment that encourages learning

2. High (but not unrealistic) expectations by parents and caregivers for their children's achievement and future careers

3. Involvement of parents/caregivers in children's education at school and in the community

When one or more of these conditions is in place, students show a number of powerful benefits: higher grades and test scores; better attendance and more accomplished homework; fewer placements in special-education classes; more positive attitudes and behavior; higher graduation rates; and greater enrollment in higher education.

RESEARCH SHOWS THAT PARENTS WHO ARE INVOLVED WITH THEIR CHILDREN...

- Have more confidence in their own ability to help their children learn at home

- Have more confidence in their children's schools

- Have higher expectations of their children

- Receive increased respect from teachers

- Are more likely to enroll in continuing education to advance their own schooling

Parent involvement clearly benefits individual students; it can also be an enormous boost for entire schools and the student body as a whole. (For more about this, see "Your involvement at school" on page 26.)

It's the successes at school that most reliably predict a child's maturing confidence and continued success in life, even beyond the content of what she learns in class. Going to school is a child's "job." When children are successful in school, they not only acquire important knowledge and skills, but build a positive sense of themselves. When you support your child's schooling, you support her sense of self far into the future. Research described by Laurence Steinberg in *Beyond the Classroom: Why School Reform Has Failed and What Parents Need to Do* (Touchstone, 1997) indicates that the school experience has a lasting and powerful impact on the growth of self-esteem—maybe even more decisive than whether or not a child has a strong sense of herself as she enters school. Of course it's true that children with strong inherent self-esteem may do better in

school. A confident child makes a confident student. She may ask more questions, be more resilient, or take on more academic difficulty. But Steinberg observes that children really blossom as they achieve academic success *during* their time in school, as they work hard and then improve— whether or not they came into the system with strong self-esteem. A child's success in school leads to a strong sense of confidence.

The research also shows that, as a parent, *your attitude* about school and learning, your *expectation* of excellence, is hugely influential on your child. The desire for excellence (only sometimes represented by grades) is very important. We discuss this in more detail under "Expect the best from your child" on page 12.

Simply, then, the ways in which you support and enrich your child's schooling today profoundly affect her lifelong education.

Involvement means engaging with, caring for, and believing in your child

The good news is that being involved comes quite naturally to most parents because it's about what you already do instinctively: care deeply for and about your child. It doesn't require a college degree, or knowledge of math, science, or any other subject in particular. It doesn't take extensive preparation. It doesn't require stay-at-home parents, or fluency in English, or a two-income family. **Research shows that a collection of concrete actions, many of them small, many that take very little or no extra time, can make a huge difference for children.**

AREAS IN WHICH RESEARCH SHOWS THAT CONCRETE ACTIONS STRONGLY BENEFIT YOUR CHILD'S ACADEMIC SUCCESS

- Your home environment
- The expectations you have for your child and the support you provide
- The attitudes you hold
- The modeling you naturally do by how you go about your own life
- The ways you engage with your child related to schoolwork
- The opportunities you provide for learning beyond school
- The ways you support your child's teacher and school

Many of these ways of being involved relate to how we go about the routines of daily life, and how we think about and approach the situations we encounter. You may be surprised at how many of these actions are already part of what you do every day.

Take a moment to think about ways in which you already support your child's education, and look at how research says they help:

HOW YOUR ACTIONS ARE SUPPORTED BY RESEARCH

- Do you have a daily routine at home...get your child ready for school in the morning...have a firm bedtime...assign chores? From a home routine, your child learns about organizing and sharing her time, and about priorities, responsibility, and self-discipline. You've established a great blueprint for how she can manage and learn at school.

- Do you have dinner together most nights? Discuss the day's events? From this kind of time with you, your child is learning to reflect, communicate with adults, and use critical-thinking skills to evaluate his day. He also sees that his efforts at school are important to you.

- Do you recognize your child's special talents...enroll her in sports or music or art programs...make room on the kitchen table for her drawing or go out and throw a ball with her? This kind of involvement supports your child's whole education, of which schooling is just one part. With this kind of support she can expand her talents, experiment freely, and keep a healthy balance of skills and interests. By playing organized sports or learning to play an instrument, a child can learn a lot about learning. These activities instill the value of practice, recognition that hard work pays off, and the knowledge that you have to be willing to work on challenges to achieve your goals.

- Do you show pride in your child's accomplishments and tell relatives and friends? This kind of attention communicates that you value your child's hard work and so does the broader community.

You're already doing things, right now, that support your child's education.

How home life prepares your child for success

Research has shown that children who do well in school have several characteristics in common at home. These similarities aren't related to money, ethnicity, or family makeup, but to—you guessed it—parent involvement. Here are some that research has found to be most effective:

RESEARCH-BASED AND EFFECTIVE
PARENT-INVOLVEMENT PRACTICES

- Maintain a warm and supportive home.

- Have a daily home routine that helps your child prepare to learn at school. Help your child get ready for school every morning, and be firm about bedtimes.

- Model the values of learning, self-sufficiency, and hard work at home.

- Take advantage of community resources for family needs.

- Have a home life that includes reading, writing, and discussions among family members, and make sure your child has regular opportunities to talk with adults.

- Share meals and discuss the day together as often as possible.

- Express realistic but high expectations for your child's achievement. Pay attention to what's happening at school, discuss possible future occupations, and encourage your child's growth and success.

- Get involved with the school's instructional program. This improves your child's attitudes about school as well as the overall culture of the school. (See "Your involvement at school" on page 26.)

- Make yourself a partner in your child's education by working with teachers, schools, and the community.

- Arrange for after-school care and supervised activities. Monitor out-of-school activities. Limit television time, and encourage your child to socialize, be active, and play.

- Create an environment of non-perfectionism (for the children and the parents in the family!).

Other Key Suggestions for Involvement

Expect the best from your child

Have and convey the **highest expectations** for your child's academic success. If the message your child gets—because you're afraid to aim "too high," or you sense her reluctance—is that it's okay to do okay, what more should she expect of herself than "okay"?

Without sacrificing realistic expectation, there's every reason in the world to convey to your child that her work level at school should be high. Make it clear that her academic success matters hugely. It does. Whether you yourself were an A or a D student doesn't matter a bit when it comes to your hopes and expectations for your child. Make every syllable of your spoken and unspoken messages about school and education shine with high expectations. Talk about her future career and the full range of possibilities open to her. Don't set the goal line too short. Your child can do well and must strive to do well; that message is worth a lifetime of remedial reassurances.

This doesn't mean holding your child to standards without room for compassion and support when she falls short. It means that your mutual aspirations should be high. It'll make her proud, and it'll give her confidence that YOU have confidence in her. And she's apt to make it happen!

Remember that how parents react to academic success or failure is very important. When parents communicate to a child that his academic successes (both small and large) are due to *his own* hard work and diligence, and that his academic difficulties stem from not working hard enough, the student learns that *he* controls his scholastic fate—not his teachers, his parents, his genes, or the luck of the draw. Like everything else in parenting, this can be tricky; some children are enormously sensitive to even the slightest criticism; others (especially as they enter their teens!) may react with rebellion and defiance to any comment, no matter how constructively intended. But do persevere, with sensitivity and courage, because it's one of the most important and long-lasting lessons you can impart to your child.

Perhaps most importantly, give your child the space to find her own internal educational compass. The mother who blurs the boundaries between herself and her daughter by assuming that because *she* wasn't motivated in math, her daughter will feel the same way, is creating obstacles and limiting her daughter's range of possibility. Same for the father who expresses disappointment when his son would rather be on the debating team or in the poetry club than go out for football. Children, though marvelously resilient, can also be very fragile.

Scottish poet Hugh MacDiarmid wrote of the unconditional love of a poverty-stricken mother for her children, "not seeking to smother any jet of their spirits in her own preconceptions or wishes." Your child is his own person, and will rise to his greatest heights if you support him as an individual.

Avoid the grade trap

It's the learning that's important, not the grades. While grades may *help* you and your child assess what she's learned, they're not always the best indicator. Some children respond positively to the incentive of getting top grades, but others suffer under the pressure and competition that comes with the grading system. Helping your child learn for the sake of self-improvement, and for the joy of learning itself, is a lifelong gift. It paves the way for positive learning experiences and successful behavior for the rest of her life. Avoid comparing your child's grades to those of her siblings or other students; it's almost always counterproductive. Instead, compare your child's report card to her last one. Ask questions, things like: Where have you improved? Where do you think more improvement is needed? Would you like to set a personal goal for your next report card? How can I help you meet that goal?

Stay involved after elementary school

The benefits of your involvement as a parent aren't limited to early childhood or the elementary level; you continue to have a positive effect through (and beyond!) high school.

In some ways, young people need your engagement **more than ever** as they traverse middle school and high school—but they need it in different ways. While some middle school teachers may welcome, for example, parent volunteers in the classroom (especially during a science event, where more adult facilitation is helpful), for the most part classroom volunteering isn't the kind of parental involvement that happens in middle school. This is true for many reasons—not least because the vast majority of our emerging self-conscious adolescents, at least when in the company of their middle school peers, find their parents hopeless embarrassments at best! And speaking of peers, the friends our children make and "hang" with become, at this age, an increasingly important part of their larger education, and can exercise enormous positive or negative influence. (You can capitalize on peer influence; see the box on page 20.)

Academically, progress through middle school and high school in our society is crucially important to our children's higher education and future careers. As a parent or caregiver, you need to become a well-informed advocate in order to help your increasingly self-reliant teenager navigate the system and make good educational decisions. Become aware of academic course requirements and prerequisites. Get involved in school-board or community decisions or campaigns involving educational improvements and changes. Your involvement at these higher stages is very important.

Consider mathematics, champion algebra!

The subject of mathematics, and algebra in particular, is worth special mention. Many studies have shown that math is a filter; completion of math courses strongly correlates to whether a student goes on to take more advanced courses in high school and college. **Competence in math is a primary indicator of a student's career potential and achievement.** In this sense, math takes on far greater social, economic, and political importance than many people realize.

Algebra is a truly fascinating branch of mathematics—one of the most amazing inventions of the human mind! When introduced in a positive way and in an interesting context, algebra can draw out the creative, problem-solving, and "puzzle-figuring-out" abilities of all students. Unfortunately, for a variety of reasons algebra often gets a bad rap, a cultural reputation for being complicated, difficult, and generally for "smart" students. (It doesn't help, that, despite excellent efforts by math educators, there's a severe shortage of qualified math teachers. Science teachers too.)

Algebra is one of the primary "gatekeepers" in tracking students along academic or non-academic pathways. Students who don't complete an algebra course by eighth grade just won't have time to complete the full complement of math courses in their four years of high school. This is an obstacle to entrance at many colleges. "Gatekeeper" courses have had a discriminatory impact on historically underrepresented racial and cultural groups—and, due to gender stereotyping, a profoundly negative impact on girls and young women. (The award-winning 1988 Warner Bros. film *Stand and Deliver* makes this point powerfully, specifically in regard to Chicano students in Los Angeles.) Many groups, including The Algebra Project, EQUALS/FAMILY MATH, and other community and parent organizations (see Appendix B) have developed ways to effectively address these issues. In many communities, parents are working with educators to ensure equal access to high-quality mathematics learning for all students.

You can be one of them.

Become a conscious advocate

Middle school and high school pose many challenges for all concerned—students, parents, teachers, administrators, and the community at large. For the student, the pace of physical, emotional, and intellectual development is rapid, uneven, frustrating, and exhilarating. The academic challenges can be intense, the homework assignments massive. For the parent or responsible adult, finding the constructive balance between guidance and independence, discipline and self-reliance, academic achievement and human growth can be very difficult, and seems to shift literally hour by hormonal hour. For teachers, middle school and high school can be a very exciting time, as their students' maturing young minds become capable of grasping and debating more complex ideas in many fields. At the same time, the social realities of today's communities and classrooms, limited resources for teacher training, and even the natural exuberance of adolescence can make classroom management an exhausting burn-out factor of major proportions. For a school administrator, there are immense pressures involved in trying to lead a non-stop institution, ensure safety, prevent violence, maintain order, resolve conflict, coordinate a complex set of systems, and strive for higher standards and improved test scores amidst declining resources. And in the community, adolescents need safe places to go and constructive things to do outside of home or school. Many communities offer sports leagues, social centers, Big Brother/Big Sister programs, and other safe, healthy, engaging options that can keep young people out of the kinds of trouble that arise from disadvantaged social conditions, boredom, or a need for belonging.

All these factors only make the need for parental involvement at these grade levels that much more important.

WAYS TO SUPPORT YOUR CHILD THROUGH MIDDLE SCHOOL AND HIGH SCHOOL

- Offer steady encouragement.

- Stay aware of course requirements.

- Pay careful attention to pathways to higher education.

- Stay in consistent communication with teachers and administrators.

- Intervene directly if serious problems develop with a teacher, a particular subject, or your child's behavior at or regarding school.

- Find safe places and engaging activities in the community for your child.

- Be a conscious advocate on many levels.

These are all components of effective parenting, as our children grow into young adults. No one person can do it all, but we owe it to our children—and all young people—to help in the best ways we can.

Connect with your child's teachers

James Comer, professor of child psychiatry and associate dean of Yale Medical School, contributed to our understanding of the parent/teacher/child relationship and created a highly successful profile for child motivation. As Dr. Comer put it, "Children learn best from the people they bond to."

The more you encourage and support the bond between your child and her teachers, the greater her trust and her likelihood to learn well from them. Make it a point to nurture those relationships. Create and sustain real partnerships of your own with your child's teachers, too, so your child sees what and whom you value.

THINGS THAT HELP BUILD A CONNECTION WITH YOUR CHILD'S TEACHERS

- Respect and understand a teacher's role in your child's education. Convey that respect to your child and her teachers.

- Every teacher has a strength and a passion. Help your children know what their teacher's gift to them can be. Remember that your child will be exposed to many teachers over the course of her education, and knowing how to find the positive in each is a healthy and effective approach to schooling (and life!).

- Invite your child's teacher to your home. Share things about your life, talk about school, ask about her family (or invite them too). Let your child see you form and sustain a respectful relationship with his teachers.

- Stay in touch with teachers and staff. Use whatever works—notes, phone calls, e-mail, or hallway conversations. Some teachers will be more communicative than others, but do your part to keep the communication flowing.

- Ask teachers what kind of support they could use most, both in the classroom and in how you interact with your child regarding homework.

Getting to know your child's teacher can give you a good sense of what your child will be doing during the school year and the kind of person she'll be interacting with each school day. It's also your chance to learn something about the teacher's educational philosophy and style. Communicating with the teacher early in the year may uncover ways in which you can help over the course of the term. And if it happens that you need to speak with the teacher later in the year about something that concerns you, the ice will have been broken. Establishing this relationship can also help you coach your child in dealing with the teacher should problems arise between them.

In our increasingly electronic age, effective use of e-mail has become a good means of communication between teacher and parent (if the teacher is amenable to it). Perhaps especially in the middle school and high school settings—and particularly when a parent has met a teacher at least once, say at a Back-to-School Night—occasional e-mail communication can help bridge the gap between teachers and parents at higher grades.

Be positive about the school experience (even when it's problematic!)

Attitude is something we pass along to our children without even being aware of it. Research has demonstrated that there's a correlation between parents' and children's attitudes about a teacher, a school, or the school experience. (Conduct a quick survey in your head of parents and children you know; doesn't this seem to hold true?)

So, what's shaped your attitudes? Was your academic experience at school pure pleasure? What luck (and how rare); conveying that glow to your child won't be a problem. Was school part-joy, part-anguish, as it was for most of us? Without denying the tough parts, concentrate on the good. That's what got you through—and more importantly, it's what made you keep wanting to learn and to excel. Are your school memories buried under years of effort to forget? You'll have to actively concentrate on the desire for—and, truthfully, the likelihood of—a different and far better experience for your child. Work to convey *that* message. If much of your school experience was rotten but some good came of it, either at the time or years later, you can consciously opt for the approach of the parent whose academic years were a mixed but decent bag of goods; you can concentrate on the positive results.

It's a lot better for a child and his success to feel positive about his own situation. School is challenging enough without extra negative baggage. So even if you have some reservations about your child's teacher, or your own school experience wasn't positive, do your best to muster a positive attitude about your child's school, homework, and teacher. If you have concerns about something in your child's school experience, work in a positive way **directly with the teacher** to address them. It's amazing how many parents criticize a teacher's practices to the child but never to the teacher! All that does is pass along a negative attitude to the child without helping to improve the situation.

> ## HOW TO KEEP THE SCHOOL EXPERIENCE POSITIVE FOR YOUR CHILD
>
> - Have a positive attitude about your child's school, homework, and teachers.
>
> - Talk directly with your child's teacher about concerns you may have, or to establish shared expectations for your child.
>
> - Approach a problem situation with a tone of concern, rather than anger or frustration (even if you feel that!). You'll be surprised at the number of teachers who are eager to work with you to change a situation, and happy for your involvement and support. Imagine how overwhelming it can feel for a teacher who has thirty students needing thirty different things. In this situation, teachers—especially inexperienced teachers—can sometimes resort to a sort of "crop-dusting" approach, providing the same remedy to all students regardless of their individual needs. Raising your hand to point something out may be just the help a teacher needs to come through with strategies that work for your child.
>
> - Laugh with your child about less-than-perfect teachers or school situations. Let your child know that as she navigates the world she'll encounter all kinds of people and situations, and can weather them just fine.

Watch for signs of reluctance

For the parent of, say, a fourth-grader named Diana who's recently started saying she "hates" math or science, every minute counts. Unless Diana can be shown NOW how exciting and involving math and science can be—and how fully capable of them she is—she may not get a second chance. A life-long pattern of avoidance and lack of self-confidence can set in that could keep Diana from ever realizing her full potential in math and science. That wouldn't just be a shame; it would put a real-world obstacle in her path for years to come. **Research shows that students who opt out of math or science as early as their middle school years drastically limit their career choices later in life**. As you saw in "Consider mathematics, champion algebra!" on page 15, key decisions about mathematics in the middle school years are absolutely crucial.

Turning your child's resistance around has its rough moments. Frustration and anger can erupt in a child struggling with a subject he doesn't understand (and therefore fears). It's important to know that these reactions often precede a learning breakthrough. Your positive attitude, your involvement, and your support can guide your child from anxiety or dislike to a confidence for trying new things. She may not grow up to be a scientist or mathematician, but she'll overcome her fear of these subjects and acquire sufficient scientific literacy to make informed citizenship decisions.

WHAT TO DO IF YOUR CHILD SHOWS DISLIKE OF OR ANXIETY ABOUT MATH OR SCIENCE

- Talk with your child about exactly what she dislikes.

- Talk with your child's teacher and ask for suggestions for dealing with the situation.

- Consider getting a tutor for your child (again, her teacher can probably help).

- Provide subject-related, anxiety-free opportunities for your child outside of school: a trip to a science museum, nature walks, a home science kit, a math game, fun Internet sites, or an informal project where math is needed (like keeping track of baseball statistics). See "Resources" on page 138 for more ideas. Find ways to point out how these unintimidating activities relate to science and/or math.

- If peers are very important to your child because of his age or personality, encourage science- and math-related activities outside of school with peers who are enthusiastic about these subjects.

- If you, too, happen to have anxiety about math or science (or even if you don't), see if an adult friend or relative could take on the role of math or science mentor. As you've no doubt discovered, children are often more open to new things when they're not with their parents!

- Remember to be encouraging but sensitive about the progress your child makes and the risks he takes. It's hard to confront one's anxieties and shortcomings.

School, Home, Community, and You

Schooling versus education: different?

Many people think of "school" and "schooling" as comprising a child's entire education. As we generally define it, schooling is finite and typically "contained." We go from elementary to middle to high school, and many of us continue to university or community college. **Education**, on the other hand, is lifelong. It happens in any space, at any time, in any company. It's all the knowledge we acquire over the course of a lifetime. And while schooling is a real and vital part of it, education itself knows no bounds. It's only natural that our early approach to education has legacies far into adulthood.

Many sources of input contribute to a child's total education: community services, extracurricular activities, the modeling you do as a parent as you live your life, other role models your child has, television, the Internet, peers, communicating openly with your child about values, and more. **Everything we've learned through research shows that a parent's active management of a child's whole education is crucial.**

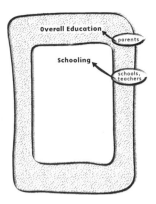

Schooling is only a part of a child's education, but it forms an important core and has a deep impact on the future. All suggestions in this book, even "school-specific" recommendations, lay the groundwork for your child's education as a whole. Everything you do to support your child's academic success does the same.

Schooling versus Education

Who's in charge of your child's education?

You are—until you hand that responsibility off to your child (a process that begins early but takes decades!). As a parent, it's your role to facilitate and support your child's total education in many ways. There's no one else playing that role. Teachers and schools are in charge of your child's schooling, but research shows time and again that they cannot do a great job of it without your involvement. It's the *combination* of home and school environments that creates what children need to succeed. It takes deliberate choice to become a partner in and advocate for your child's education, and your commitment is the only way it can work. Happily, getting your mind to "go there" gets easier very quickly. And, naturally, the more you commit to involvement, the more second-nature it becomes.

Bringing Schoolwork Alive

by Barbara Nagle
Co-Director, Science Education for Public Understanding Program (SEPUP)

What's your child doing in school this year?

When our two daughters (now teenagers!) were little, we used their school studies as springboards for exploration. We called the weekends "family days," as a contrast to weekday "school days"; and for at least one of those days we explored. We made a point of getting acquainted with our part of the world and enjoying activities that were meaningful to the whole family.

Our closest metropolis is San Francisco. Many of our family days involved outings to "the city," discovering free or inexpensive learning opportunities among the endless resources a big city offers. When the girls were studying Egypt, we went to the Egyptian exhibit at the art museum. When they had to write a report about sea creatures, we visited the aquarium.

We learned that the best way to interest our children in history was to put them right in the middle of it. When one of our daughters was learning about California history, we searched for nearby historical sites. Within twenty miles of our home we discovered a turn-of-the-century estate, a Victorian farm, and a reconstructed Ohlone Indian village. For our children, as they stood in or beside such tangible historical symbols, history took on real and immediate qualities.

Our daughters' understanding of the world also grew as we explored interests they developed outside of school. Trains, for instance, have always been a passion in our family. We've been on every train we can find—including subway trains!—in the San Francisco Bay Area. For a family lesson in who and what helped make our state what it is today, we took the train (naturally) to the California State Railway Museum in Sacramento. Walking the length of the gargantuan Big Boy locomotive and the new Amtrak passenger cars, reading the many posters and listening to the museum's film presentation and "talking dioramas," our daughters learned firsthand about the role of the railroad and the great diversity of cultures that went into building our state.

So, what is your child doing in school this year? Is your son studying ecology? A trip to a local botanical garden would allow him to imagine himself in exotic ecosystems. Is your daughter taking biology? A trip to the zoo would let her apply her new learning. Architecture or city history? We discovered and toured buildings designed by renowned California architect Julia Morgan.

Other outings take advantage of special local events. We discovered free university "open houses," which gave our children the chance to enjoy chemistry demonstrations, a tour of a research lab, and a visit to an anthropology museum. Our city holds free multicultural fairs and farmers' markets on the weekends, where we enjoy the sights, sounds, and aromas of other traditions. Local libraries hold reading nights. Our YMCA offers inexpensive dance lessons that explore the performing arts of many cultures. In many communities the local recycling center has an annual Earth Day celebration, with music, art, and demonstrations that teach conservation. Local regional parks offer free walking tours organized around historical, geological, or ecological themes.

These activities near our home have provided our daughters with a frame of reference that helps them more deeply understand school

> *subjects such as science, history, and the arts. When one of our daughters was young, several of her teachers commented on her unusual understanding of the world around her as a result of these local excursions. Every place we've lived in or visited has had numerous activities and sights just a short walk, bus ride, or drive away.*

Your involvement outside of school

Have a look at the overlap in Figure 1, "Home/School Intersection." (You may remember this kind of tool. It's called a "Venn diagram," after nineteenth-century English scholar John Venn—who probably didn't invent it, but that's another story.)

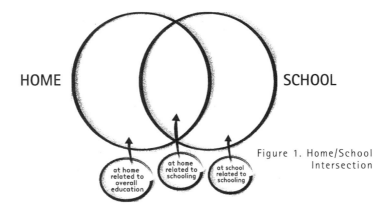

Figure 1. Home/School Intersection

Two interesting points emerge from this diagram that at some level may seem obvious but are often lost in practice.

Much of your involvement at home relates broadly to your child's overall education, as well as to her schooling. Single parents or working couples often assume that they just can't be involved in educational activities because they aren't available when school's in session. They may be tempted to relegate that domain of a child's education to the schools. It's easy to forget that, as a parent, you can choose from a vast number of out-of-school opportunities for learning that are *at least* as powerful as what happens at school. Many of these can occur while you're driving in the car with your child, cooking, on a family outing, shopping, or working in the garage.

The second point demonstrated in "Home/School Intersection" is that **much of how you support your child's schooling occurs at home**. Just as some parents assume they can't be involved in school-related activities, school personnel sometimes jump to the conclusion that single parents or working couples can't be approached or relied on for parental involvement.

But in fact there are many powerful, efficient, and focused ways in which these parents, whose time for participation at school is limited, can be involved. They absolutely can play active roles in supporting the school—and specifically, their children's schooling. Good teachers and principals have come to understand this.

One of the most time-efficient ways for a parent to support his child's education is to engage in learning activities at home that are related to what's being learned at school. Another is to read, read, read—together! It's well documented in the research that reading at home produces significant gains in student achievement. There are many other things you can do as a busy parent that make the most of your time and reap tremendous academic rewards for your child. For parents who are simply unable to make it to the school, home learning activities provided by the teacher are a great way for school and home to build on each other.

Schools and teachers are adjusting, capitalizing on what's been learned about the powerful combination of home and school, parent and teacher. As parents, we need to make that shift, too.

THINGS TO DO AT HOME THAT SUPPORT YOUR CHILD'S SCHOOLING

- Show genuine interest in your child's progress at school.

- Read to your child, and have your child read to you.

- Help with homework (see page 79 for more about this).

- Participate in home learning activities with your child—either those that come home with your child's schoolwork or other, related activities.

- Give spelling and math drills.

- Reflect with your child on both her strengths and the areas in which she needs improvement. Together, make a plan that supports your child's improvement—and then celebrate her successes!

- Stay in touch with teachers and school staff (send notes, phone, e-mail).

- With your child, watch a documentary, an educational video, or a movie that relates to what's being studied at school, and talk about it together.

- Organize or participate in a class telephone or e-mail tree to communicate important information among parents in your child's class.

- Offer to gather or purchase materials needed for class projects or science activities.

- Offer to write a feature for the school newsletter.

THINGS TO DO OUTSIDE OF SCHOOL THAT SUPPORT YOUR CHILD'S EDUCATION

- Read to your child and have him read to you. Discuss what you've read. Write letters, lists, journals, and messages together.

- Recognize and encourage special talents. If you can, enroll your child in a (reasonable) schedule of sports, dance, music, or other classes. Or just encourage her to play chess, beat a drum, or otherwise pursue her abilities.

- Take advantage of such community services as Boys and Girls Clubs, libraries, the Y, and community-center classes.

- Expose your child to many learning opportunities outside of school. Visit museums or nature centers, go to concerts or the theater (many performances are free), take advantage of community events.

- Ask and encourage questions, and use reference materials and the library to pursue interests and answers.

- Hold real conversations with your child. Listen to and talk with your child about things that are important to you both. Tell stories and share problems. Reflect on lessons learned through daily experience (including your own!).

- Talk to your child about current events; broaden his view of the world.

- Involve your child in a family research project, such as investigating your family tree or using *Consumer Reports* to choose a household appliance to purchase.

- Talk about people you admire and why. Expose your child to adults who might serve as role models or mentors.

- Discuss the value of a good education. Talk positively about the school experience.

- Have high expectations for your child's academic success. Show interest in her progress at school. Express interest in and high expectations for her education after high school, and for her career choices.

- Set goals and standards that are high but appropriate for your child's age and maturity.

- Involve your child in planning a family activity such as designing a garden, thinking through a household repair task, or planning a family trip.

- Engage your child in fun science and math activities and projects at home. (See page 147 for resources on this.)

- Inform friends and relatives about your child's successes.

- Especially as your child enters adolescence, look for other adults who can provide some of these experiences for your child.

Your involvement at school

There are good schools and there are great schools. **An involved parent body is one of the strongest indicators of a school's likelihood to be great.** Teachers' morale improves dramatically when they feel that parents strongly support them and their work. (No surprise there.) Teachers with high morale naturally bring extra gusto and care to their teaching, which has a direct impact on our children. Student achievement is higher. The school enjoys a better reputation in the community. Parents and members of the community feel increased ownership, resulting in increased support of schools and willingness to pass local tax bonds to support the school. It's a powerful cycle, and parent involvement drives the engine of its success.

As we've recognized, parents are not equally available to be actively involved at school. And every parent has times (or years) when he is more or less available. Resist the temptation to generalize about others or yourself ("I have three children—I can't possibly find the time to be involved"; "She's a single parent, she probably can't contribute to the school newsletter"). Sometimes you'll be able to participate, sometimes you won't. Again, research shows that a parent's *decision* to be involved, no matter his circumstances or amount of available time, has the greatest impact on a child's education. Studies have found that parent involvement occurs *not* just in high-resource, leisure-rich communities, but equally vigorously in com-

> Parents who are guided by teachers on how to help at home spend more minutes helping their children with homework than other caregivers.
> —Dauber and Epstein, 1989

munities principally composed of working parents, low-income parents, or largely non-English-speaking parents. And in communities in which parents aren't involved, it's usually not strictly because of circumstance, but because those parents opted out or received no support from the schools. **The critical difference is not the situation of the parents, but rather the organization and support provided by the school to develop and maintain partnerships with families.**

If you're lucky, your school already has a high level of parent involvement and you can step right into the positive flow. If not, your participation can help turn the tide—and benefit your child directly. While there's some evidence that the children of parents who are involved at school are the students who benefit most, studies show that *all* children develop better attitudes about school and schoolwork when parent volunteers are in the classroom. The general presence of parents in the classroom communicates that schools and schoolwork are valued and important in the community.

Whatever the situation at your school, your participation there will help its trajectory toward becoming a great school.

There are many diverse areas in which your child's school can use your help:

- **Academics**. Be involved with the academic program as a classroom volunteer.
- **Fundraising**. Help raise funds to pay for enrichment activities, either by conducting fundraising events or writing grant proposals.
- **Materials**. Gather or help provide materials for science activities.
- **Decision making**. Serve in a decision-making or oversight role, such as being a member of the school site committee.
- **Advocacy**. Serve as a school proponent/supporter in the community, at the school board, with local businesses.
- **Community Liaison**. Help your school form partnerships with business and community organizations—starting with your own!

Every parent has different interests, talents, and available time. The key is finding your way (and your year) to be involved at your child's school.

PRACTICAL WAYS IN WHICH YOU CAN SUPPORT YOUR CHILD'S SCHOOL—AND BY EXTENSION, EVERY CHILD IN IT, INCLUDING YOURS

- Ask teachers how you can help.
- Visit your child's classroom; sit in on classroom activities or lessons.
- Share your knowledge, skills, or interests with your child's class.
- Help in the classroom. Work with students to provide extra support.
- Help with field trips and other extracurricular activities.
- Assist your child's teacher by obtaining and preparing class materials.
- Join parent/school decision-making committees.
- Learn about school programs and needs, and then advocate for those needs to your community—your neighbor, the school board, community organizations, etc.
- Raise money for school projects.
- Call the local newspaper to get news coverage for school events or unusual class projects.
- Volunteer to help office staff conduct school mailings.
- Organize community "appreciation events" for teachers, such as cooking and serving a meal on a day when teachers need to stay late, or organizing a "thank you brunch."

In the area of math and science education reform, school/parent/community advocacy sometimes arises *ad hoc*; spontaneously and out of circumstance. This is especially true when a group of parents, concerned teachers, or even active students themselves realize that math and science are critical to higher education—and that their school or school district hasn't kept up with the times. Here are some examples:

- A parent's organization that held community forums and published its own newspaper was established in a large city in the Midwest to pressure the school board to adopt inquiry-based science programs, and to work more closely with the National Science Foundation and other voices of educational reform.
- As in the example described in "The Mystery of Parent Involvement" on page 42, active parents have often lobbied and organized to encourage administrators or school boards to improve science and math education.
- The Algebra Project (see page 131), which seeks to revolutionize the

teaching of algebra, urges middle school and high school students themselves to become active organizers of improved and more relevant math instruction—not unlike the student organizations that helped launch the civil rights movement!

- The Mathematics, Engineering, and Science Achievement (MESA) organization has helped thousands of historically underrepresented students pursue careers in scientific or technical fields. In states where it's active, MESA can be an ally in these kinds of partnership.
- Associations of women in different fields of science can be a strong source of support, and can often provide role models who can discuss their experiences during classroom visits.
- Professional organizations of scientists and engineers who are Latino, African American, or Native American can also be very helpful. A group of Black engineers in Northern California, for example, has put together a street festival that can be held at schools, museums, or community gatherings, with large posters showing successful scientists as well as hands-on activities for children and their parents.

Sometimes your research or your inquiries in the community can be just the kick-start a school needs to bring its math and science curriculum up to speed.

How Do You Begin?

Start with the "comfort zone"

We hope that by now you're seeing all the ways in which you already actively support your child's education—and especially, how much of what you do every day "counts" as educational support. This is what we call your "comfort zone," all the familiar areas in which you feel at ease. This is the place to start. Feel good about what you're already doing and recognize its importance!

It's also important to remember, as you build your confidence and your repertoire of learning opportunities, that you can't possibly provide every form of involvement from which your child might benefit. In other words...

You can't do it all!

No one can do everything on all these lists. And that's not the goal—or what research says is necessary. The research shows that parents and adult caregivers who do these *kinds* of things have children who succeed academically. There's no uniform "look" to a successful situation. So the first thing is to **resist feeling that what you do doesn't measure up in some way.**

Think of your involvement with your child's education as a continuum. If you're like us, some of the things that research shows influence a child's success later in life don't strike us as all that important in the moment. But even simple efforts can have big payoffs down the road. If any of the ideas we propose throughout this book strikes you as inviting and feasible, think about starting to fold that kind of activity into your life. As you add, here and there, to things you already do, you expand your comfort zone. Soon it will just be part of the routine, part of the way you and your child interact.

We know that parenting is hard work! The rewards (and the thanks, if any) often come years later. We hope that by knowing what's been shown to make a difference you can confidently focus your energies in certain arenas. By doing the best you can to be involved, you'll know that, although the results might not yet be apparent, you're doing the right thing.

A Gardener's Bookshelf
by Deborah Lee Rose
Author and children's book reviewer

Each spring when I was growing up, the garden in front of our small row house would "magically" burst into bloom with daffodils, irises, and the rosebuds that, by summer, filled our yard with bees and blossoms.

The magician behind this floral extravaganza was my father, who also transformed the pebbly patch in back of our house into a thriving vegetable garden. There, through the hottest days, he hoed, seeded, fertilized, watered, weeded and harvested until every square inch had produced something edible.

Although I didn't inherit my father's green thumb, he bequeathed me a deep love of gardens that I want to share with my children. My seed catalog is the local library, my garden center is the bookstore, and all around our house I'm planting books that nurture children's instinctive love of growing things.

The Rose in My Garden, by Arnold and Anita Lobel, is an irresistible tale written in witty, add-on verses—a gardener's "This is the House That Jack Built," full of vibrant description and gentle repetition.

Told in a child's voice, Planting a Rainbow, by Lois Ehlert, sows flowers of every color. Ehlert's companion book, Growing Vegetable Soup, captures the simple ecstasy of harvesting one's very own fruits of the earth.

Miss Rumphius, by Barbara Cooney, cultivates a love of natural beauty. Alice Rumphius sits on her grandfather's knee and dreams of visiting faraway places and someday living by the sea. "That is all very well, little Alice," her grandfather observes, "but there is a third thing you must do. You must do something to make the world more beautiful." Much later in life, Miss Rumphius discovers how; she transforms the rocky sea coast where she lives by planting a wind-flung garden of thousands of blue, purple and rose-colored lupines....

Your family context

Our family situations never seem easy enough. Dual working parents, by definition, have limited time with their children. And they arrive home to this priority—making time for each child—*after* they've already invested a huge amount of energy at the workplace. A one-parent household takes an enormous toll on a working adult's time and energy. A large family with lots of competing needs can dramatically cut into one-on-one time with each child—and on all-round time available for educational concerns. When English isn't a family's first language, school materials may be hard or impossible to read with your child, or you and your child may not completely understand what his teachers are asking him to do. If there's a large gap between your home culture and the dominant culture at school, you may feel very "different," and uneasy interacting with teachers or other parents. This can be isolating. And if it happens that schooling wasn't possible when you were growing up, it can be intimidating to walk into a child's school, or open a current textbook, or hear about "standards" and other wordy educational matters.

Of all potential obstacles, the situation of family poverty presents the biggest challenge. When money's tight and social conditions difficult, the priority is to meet immediate real needs…intense stress can wear all members of the family down. Not surprisingly, there's data to suggest that poverty is the most reliable statistical indicator of academic failure. Our society has a long way to go in narrowing the gap (which may even be widening) between the resources of schools in poor and well-to-do communities. In *Savage Inequalities: Children in America's Schools* (HarperPerennial, 1992), renowned teacher and social activist Jonathan Kozol writes powerfully and persuasively about socioeconomic disparities in American schools. (See pages 139–142 for books by Kozol and others on this topic.) Nonetheless, parent involvement and family support, even (and maybe especially) in less advantaged situations, can make ALL THE DIFFERENCE. The other side of the coin of poverty's negative impact is that it remains true that a parent's involvement is the most reliable indicator of a child's academic *success*. This means that **a child with the advantage of an involved parent or other adult caregiver has a lot going for her, regardless of economic circumstances.** Parent involvement is *that* powerful.

Given that all the very real family circumstances mentioned above are difficult or impossible to change, it's a huge relief to remember the positive and optimistic research findings: your involvement alone can make the difference your child needs to be successful academically. Keep that positive vision when things feel hard!

It's never too early or too late to start. No matter your background, educa-tion, economic situation, or environment, your role is critical—and you're already, right now, fully equipped for it. It just takes the decision that you'll be as involved as possible every step of the way through your child's school-ing, to support and equip him for lifelong education.

Reaching Out

You're not alone

No matter how isolated your physical or emotional situation may feel, you belong to a motivated, unified group: parents and adult caregivers of school-aged children. The common thread is that every parent wants her child to be happy, to be safe, to be successful. That universal desire is a powerful force, whether you're a family of five in Tallahassee or a single father in Nairobi. Every parent wants to feel qualified to help his child—and every parent is. These desires and qualities bind you to other caregivers of young people. All parents want access to basic, practical, here-and-now steps to make a difference.

Parent-support groups abound. If you have access to the Internet, a search using key words turns up many options. If you can get to a library, the refer-ence section (and the librarians themselves) can point you to dozens more. There are books. There are videos. There are neighbors. If you have a special-needs child, the local chapter of the association that responds to those needs can put you in touch with families and schools that lend support.

One note: Ask yourself what you're looking for in a parent-support group. There are social resources for single parents, new parents, step-parents, gay and lesbian parents, widowed parents, elderly parents, and more. Some of these are at a statewide level; many others are regional or local. There's help for children with a variety of challenges—blindness, epilepsy, bedwetting, fetal alcohol syndrome, autism, Down's syndrome, and many other or much rarer conditions. Many of these are superb resources.

If you're mainly looking for academic support, or support in coping with issues of schooling and education in general, take that into account when focusing your search. In the online publication "Accessing Parent Groups," by the National Information Center for Children and Youth with Disabilities (NICHCY), deputy director Suzanne Ripley suggests several questions to ask yourself when looking for the right kind of support. These are useful for **all** parents, not just the parents of a special-needs child. Here are several questions, adapted with permission:

HOW TO FOCUS YOUR SEARCH FOR A PARENT ACADEMIC-SUPPORT GROUP

1. What are your personal needs? Do you want to know other families who have schooling questions like yours? Are you looking for moral support or advocacy training? Do you need homework support, or a play group for your two-year-old? Do you want to get involved in local decision making on school issues? Would you like to hear speakers on education topics?

2. What parent groups are available in your community? Check with the staff at local schools, preschools, or early-intervention programs, or even the social-service department at the children's hospital. Remember that it's not what a group is called that's important, but what it does.

3. What kind of a "joiner" are you? Are you comfortable with formal organizations or do you prefer less structure? How involved do you want to be? You can usually attend meetings once or twice before becoming a member. Sometimes the best way to become involved is to volunteer to take a role in some activity. By participating, you'll get to know other members and learn more about the goals and functions of the group.

4. Look at alternatives. Often there's no organized local group that meets your family's needs, particularly in smaller rural communities. Remember, you don't have to be a part of a formal group in order to reap the benefits. A simple one-to-one friendship over a cup of coffee, or a picnic with even one other family in your community, can be mutually beneficial.

You might also consider starting a focused parent-support group of your own....

Starting a parent academic-support group

The California State Department of Education describes several of the benefits and functions of an organized parent-support group. At its best, a group can provide a lot of concrete support:

BENEFITS OF AN ORGANIZED PARENT-SUPPORT GROUP

- It provides ongoing support, and
- ...helps in time of crisis
- ...reinforces positive coping behaviors
- ...helps focus discontent, and uses its group's energy in positive ways

- …shares information, ideas, and resources

- …provides training for parents to increase their skills

- …helps parents deal with relevant agencies

- …provides the opportunity for parents to form new friendships and decrease isolation

Consider starting small. You might just speak to neighborhood parents, or post a flyer, or ask your child's school to coordinate a sign-up (if they don't already know of a parent-support group). A flyer or bulletin might look something like this:

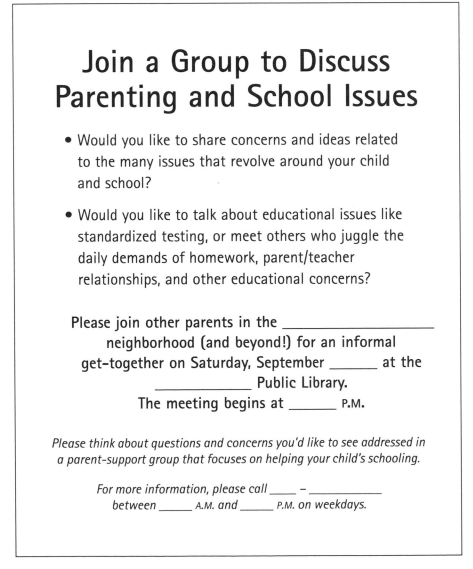

Join a Group to Discuss Parenting and School Issues

- Would you like to share concerns and ideas related to the many issues that revolve around your child and school?

- Would you like to talk about educational issues like standardized testing, or meet others who juggle the daily demands of homework, parent/teacher relationships, and other educational concerns?

Please join other parents in the _____
neighborhood (and beyond!) for an informal
get-together on Saturday, September _____ at the
_____ Public Library.
The meeting begins at _____ P.M.

Please think about questions and concerns you'd like to see addressed in a parent-support group that focuses on helping your child's schooling.

*For more information, please call _____ – _____
between _____ A.M. and _____ P.M. on weekdays.*

When Schools Are Parent-Friendly

Promoting strong partnerships between families and schools

Educational reform embraces the importance of a parent's involvement in a child's education. Increasingly, bolstered by considerable research and tangible results, we've also come to understand that partnerships among school staff, parents, and members of the community are essential in creating a thriving school community. A number of state offices advocate effective school/parent/community partnerships; here are just a few examples:

- **The Connecticut State Department of Education** co-created the "School-Family-Community Partnerships Project" to provide training, workshops, a newsletter, and a collection of resources.
- **Montana's Center for Community Education** piloted "Even Start," a project that emphasizes parents' involvement as communicators, supporters, learners, teachers, advisors, and advocates.
- **The California State Board of Education's** "Parent Involvement Policy" helps schools and school districts develop strong, comprehensive parent-involvement policies at all grade levels and in a variety of roles.
- **West Virginia's** "Total Village Project" supports parent/community/ school communication with parent-teacher action teams, mentoring, tutoring, and other assistance.
- **In Louisiana and other states**, all Title 1 schools (which receive extra educational services for children who risk failing to meet state academic standards because of poverty or other social circumstance) make written "compacts" with parents as part of a state-mandated parent-involvement policy.
- **The Colorado Department of Education** opens its Web site with just one motto: "Students-Parents-Teachers."

Such partnerships are being blazed across the country. The North Central Regional Educational Laboratory (NCREL) alone, in its Urban Parent Involvement section, lists seventy-four publications and sites on parent/school partnerships nationwide (see pages 150–151 for this and other parent-partnership resources).

Twenty-five years of research are having an impact!

The Great Explorations in Math and Science (GEMS) program at the Lawrence Hall of Science published a handbook with a variety of resources that can be used to build these partnerships. *Parent Partners: Workshops to Foster School/Home/Family Partnerships* (GEMS, 2000) is a practical handbook for educators who wish to make presentations to parents on their critical role in their children's education. Based on research and the firsthand experience of its developers, *Parent Partners* is designed for educators and activist parents seeking to increase the parental involvement in their school community. It includes step-by-step instructions for presenting parent-education sessions focused on a variety of educational topics; take-home handouts for parents and other caregivers; concrete, practical things to do to help children succeed in school; and short digests of research findings that can be used in school newsletters to motivate parents to get involved.

Veteran educator Michael G. Fullan, in *The New Meaning of Educational Change* (Teachers College Press, 2001), argues strongly for parental involvement in the schools. He places a lot of responsibility on the schools and school districts themselves to make this happen, urging them to create an environment that encourages and is responsive to parents' participation. There are concrete levels at which schools must strive to involve parents...

HOW SCHOOLS CAN CREATE AN ENVIRONMENT THAT SUPPORTS PARENT INVOLVEMENT

- At the school/classroom levels: Make parental involvement an embedded, fundamental part of the academic structure, not a courtesy.

- At the teacher level: Provide teachers with materials that are accessible to parents, to make parents' participation meaningful.

- At the program–assessment level: Actively involve parents in the appraisal of new activities and programs. Bring them in when setting objectives and make them part of the monitoring system.

- At the staffing level: Create an official, permanent staff position (even if part-time) dedicated to coordinating communication and interaction among teachers, the principal, and parents—all year, every year.

- At the school/district levels: Infuse policies and programs with parent-involvement components. Provide roles for parents in policy-making from the very beginning. Make parental contribution a cornerstone in the development of new school programs.

Partnership success stories

These vignettes describe some of the many successful parent-involvement models across the country. If ever you doubted that a "culture of engagement" with your child's teachers and schools can make a difference...please read:

PARENTS AND SCIENCE: THE TRANSFORMATION OF A SCHOOL

Science just wasn't being taught in this Peninsula school; a K–5 elementary school in Northern California. Teachers were beleaguered with the demands of teaching students to read and write, and with preparing them for the statewide standardized tests on reading and math. Science was last on the list of priorities—especially the kind of science experiences that required more than reading.

Carol, a parent who loved science, made sure to present several exciting science activities each year in her child's K–3 classes. In Carol's case, the activities came from the Great Explorations in Math and Science (GEMS) program, a collection of supplementary science units for preschool through 8th-grade students. (The GEMS units have step-by-step instructions, so they're easy to present even if a person doesn't have much background in science or teaching.)

The school's teachers were thrilled to have the help, and the children adored the activities and hands-on materials. Each year Carol worked with her child's teacher to decide what GEMS unit to present, making sure it related to the district-wide standards.

As her second child came along through the same grades, experiencing the same kinds of success, Carol decided to try to institutionalize this approach in the school curriculum. She got together with another activist parent and fleshed out the idea. They consulted with the principal, presented the idea at a school staff meeting, and got more input from the teachers. Their suggestions were adopted! The school agreed to make this parent-assisted science enrichment program a regular part of every school year.

With the input of teachers, three GEMS units were identified for every grade level, K through 5. The PTA purchased or assembled kits for those units, and the kits are now stored in a closet at the school. Every year, Carol puts out a call for parents who've expressed interest in presenting the GEMS science units. She and another activist parent hold "training sessions," modeling how to use the kit to present the unit to students. In the training, parents actually go through the unit as the children would. As parents gain experience and confidence with specific GEMS units, they share tips about what works really well, collect additional materials to enhance the kits, and volunteer to assist with future training sessions. With this energy and dedication, the program has flourished and gained momentum over time.

In the case of this story, it took only one parent to help turn science learning around—not just for her children, but for every student in the school. And after several successful years of parent participation in the science enrichment program, the school hired Carol, the parent who'd originated the idea, to coordinate the program for the whole school.

Partnerships are launched from *within* schools and school districts, too. When a teacher or administrator takes real action to make partners out of parents, and those parents jump on board, their combined forces can be awesome!

THE LEGACY OF MCKINLEY NASH

McKinley Nash persevered.

The schools in Inglewood, California, an inner-ring suburb of Los Angeles, face many challenges typical of urban education: few resources, stressed and stretched families, disconnection between the schools and the community. When McKinley Nash became Superintendent of the Inglewood Unified School District in 1994, the district was teeming with problems and suffering from low morale.

Nash thought he knew of at least one way of addressing these issues. In his previous job, as Superintendent of the Centinela School District, Nash had instituted "Quality Circles," a process used in business to help employees give input to managers. Adapted for use in schools, the system encouraged teachers and staff to let administrators know what was wrong within the system and suggest ways of fixing it.

While working for the Association of California School Administrators, Nash met Stephen Thom, a mediator for the U.S. Department of Justice who had extensive experience with "study circles." Nash immediately saw the potential of study circles to bring the principles of quality circles to bear in engaging parents and other community members. He mandated that the schools in his district organize at least one study circle apiece.

After several months of planning, twenty-eight study circles were held at twenty Inglewood schools in late 1997 and early 1998. Most of the principals and all of the school's community liaisons were trained as study-circle facilitators by Stephen Thom. Nash was determined to have the principals go through the training—in part, so they could strengthen their listening skills.

Bringing Parents into the Loop

Educators immediately noticed that the study circles helped improve relations between Hispanic and African American parents. Inglewood's students are roughly sixty-percent Hispanic and forty-percent African American, and the schools had experienced violence between the two groups almost every year during the time of Cinco de Mayo, the Mexican day of celebration. "Before," said Sophia Dossman, one of the community liaisons, "there wasn't much interaction between Hispanic and African American parents. The study-circles process, and the cultural sharing that happened during the process at my school, helped form more positive relationships."

In addition to improvements in parent relations, the overall level of parent involvement has risen since the development of study circles. Participation in PTA meetings, donations to schools, and volunteer participation at schools have all increased. Some staff community liaisons have learned to refine and tailor parent-involvement opportunities more than in the past. One liaison, for example, realized that the parents at her school work long hours and many different shifts. When she enlisted parents' help in painting school buildings, she asked one set of parents to do the preparation so that the following group could come in and do the painting itself.

The new partnership between parents and schools is evident in many new projects at Inglewood schools. Many parents now volunteer their time to clean school facilities and take care of the shrubbery and school gardens. Across the district, after-school programs and community activities (such as ESL and computer classes for parents) have also been implemented.

The schools have gained credibility by following through on many of the recommendations made by the study circles. At one school, parents suggested that the principal needed an assistant; Nash gave the approval to hire one. At the same school, study-circle participants recommended a new team-teaching strategy; it's now in effect. The circles have also helped individual schools become more confident in stating their desires to the district administration. The study-circle report to the district from Albert Monroe Magnet Middle School reads: "The parents feel that their time involved in this process should not be taken for granted, and that some movement on their suggestions should take place in a timely manner."

Empowering the Community

Exciting as they are, all of these school-level outcomes pale by comparison to what happened at the community level. For some time, Inglewood had been experiencing a shortage of funds for capital improvements. In many schools the roofs leaked, asbestos was still present, and space was so limited that libraries and science labs had been converted into classrooms. In the study circles, participants not only became convinced of the need to improve school facilities, they also gave substantial input on what those improvements should be. They rallied to pass a bond issue called Measure K; parents and students volunteered time at campaign headquarters, created phone trees, sent out mass mailings, and set up various means of outreach to educate others about the measure. Measure K passed with eighty-eight percent of the voters in favor.

McKinley Nash will be remembered not only for his voice, but for helping so many others—parents, schools, and members of the community—find theirs.

Adapted with permission from the Study Circles Resource Center (www.studycircles.org.)

"HEY! I WROTE ALL THAT?

Educators, teachers, parents, and other community volunteers have teamed up to participate in a hugely popular "Writers' Room" program at a high school in Berkeley, California. Berkeley resident Mary Lee Cole, an educational program designer, introduced the idea, adapted from one a New Jersey school district has used for nearly ten years to help narrow student achievement gaps.

In March 2001 Cole trained more than fifty volunteer "writing coaches"—adults from all walks of life—to work one-on-one with Berkeley High School students once a week. In the first three months, the coaches worked with some 300 students at the school, most of them freshman.

An article in the *Berkeley Daily Planet* (June 11, 2001) records the moment when the chairperson of the English Department saw "this neatly dressed cluster of nurses, accountants, carpenters, screenwriters, and other professionals, waiting patiently outside her classroom, wanting nothing more than to teach her students how to write." At first, many of the students didn't want to go with the writing coaches because they didn't know what to expect. Once initial hesitation was overcome, however, the program turned out to be an enormous success. And since a coaching team works with *all* students in a class, no youngster feels singled out for "improvement."

The Writers' Room program provides an opportunity to customize a key part of the education experience to the needs of individual students. By sitting down with students and helping them work through writing assignments detail by detail, the coaches offer the kind of support that keeps students from giving up. The coaches, by engaging in casual conversation on an assigned topic, often show a student that he really does have a lot to say. They help students get those difficult first sentences down on paper, and then a few more sentences, until students realize—"Hey! Look what I wrote!" One teacher said she'd seen some students go from ignoring assignments altogether to turning in neatly typed essays—a great indication of new confidence in their writing abilities. The program offers nonjudgmental interaction and a calm environment where students can focus, away from peer pressures and classroom acting-out.

The Writers' Room has already narrowed the achievement gap at Berkeley High. One English teacher said her students went up an average of one full letter grade after working with the writing coaches. The program's popularity is on the rise. By next year the program expects 200 coaches volunteering an estimated 9,000 hours—enough to make coaches available to all of Berkeley High's 900 freshman and several 10th- and 11th-grade classes. A pilot program is now planned for middle school.

Students at Berkeley High School have been given a chance to learn, hone, and use writing skills on which they might otherwise have given up. Improved writing skills are a great help in all subject areas, and an essential, lifelong asset.

The Reluctant Writer

by Katharine Barrett
Director of Family Health; School Programs, Special Projects

Enticing my eight-year-old son to read and write was a constant challenge. Wade loved to explore the backyard for insects and salamanders, and like most of his friends, he had a passion for computer games and action figures. I capitalized on these interests, and his love of adventure, to begin regular journal-writing sessions together. One winter day, snuggled up on his top bunk bed, we both set out to write for ten minutes. One foggy evening we sat in the back seat of the car with the windows open, listening to the chorus of tree frogs. Each day I encouraged him to pick a new and interesting place for us to write. He didn't think I'd take him out on the roof to write, but I did.

At first Wade rarely wrote more than two sentences—but those small steps were very enjoyable for us both. The journal entries became longer and richer with time. I found that his natural curiosity about the environment and environmental sensations were compelling for him to put into words, and we often had great conversations days later about what we'd seen from the window or roof. My daughter, three years younger than Wade, observed our fun and insisted on starting her own journal and choosing her own special spots for us to write. I suspect Heather's continuing love of journalism stems from those first sessions as an avid five-year-old author.

Nature books and field guides also provided great incentives for my children's projects and reports. Although they couldn't read much of the text, they pored over the star guide as we observed Orion in the night sky. And we all learned to classify cloud patterns from the weather guide. To this day, as successful high school students, Wade and Heather still use the bird, insect, spider, star, and rock guides we purchased for them as children.

As my son struggled to use metaphor and simile in his English assignments, we observed his terrarium and fish tank, finding poetic parallels in nature. Mapping the location of webs for a science project, he discovered that ten different kinds of spiders lived on the outside of our house!

Sometimes, when my teenagers are discouraged over a school assignment, I pull out an old journal or report and we chuckle about spelling and grammar. These simple views of progress always boost their self-esteem, and bring back good memories of learning.

2-8-93

We are Writing from the back
Seat of the car. We are coverd with
Sleeping bags. The night air is fresh and
clean. And we are amased of the
frogs we hear. It is vary nise out
hear. We are chowing gum and it
Tasts good!!

THE END!!

7:58 p.m.   February 4, 1993

Benji is curled up in his bed
looking with wondering eyes
at Wade and me. We are
sitting on the floor of Wade's
tent, writing by flashlight.
The week has galloped by.

And when entire communities get involved in learning partnerships, the results can be magic:

THE MYSTERY OF PARENT INVOLVEMENT

For some years, the Lawrence Hall of Science has designed successful school outreach programs to bring exciting science activities to schools throughout Northern California and beyond. One of the most successful of these is the "Mystery Festival," an all-school event in which students engage in forensic science to try to solve a fictional crime. The festival includes a crime scene with numerous clues, and stations at which tests (such as fingerprinting, powder analysis, chromatography) are conducted in an attempt to determine "who dunnit." It's intriguing and challenging—and it teaches a whole lot of solid science.

When the Great Explorations in Math and Science (GEMS) program adapted the mystery idea into a teacher's guide for the classroom (with optional suggestions for an all-school program), we expected it would be popular—and it is! Many teachers have set up the crime scene and stations in their classrooms, and we constantly receive letters from students claiming to have the solution! Students throughout the country have grappled with complex logic as they've worked their way through the mystery. (A Pennsylvania teacher's use of the activities was featured on the front-page of *Education Week.*)

There are two versions of the mystery: one for younger students, called "Who Borrowed Mr. Bear?" and one for older students, "The Case of the Missing Millionaire" (also known as "the Felix Mystery"; it centers on a character named...uhm...Felix Navidad). He's joined by a number of other cleverly named suspects. For the older mystery, there's no one "right" solution—which can be initially frustrating for students, but means that discussion and debate of clues and theories continues long after the activity.

We expected Mystery Festival would be a success, but we didn't realize what a fantastic vehicle it could be for parent involvement and community education until we started to hear from activist parents who were organizing all-school gatherings. Some of these parent groups were using the activities to help educate other parents, teachers, and the community about the educational value of activity-based science. One of the most successful efforts of this type took place in Ridgebury, Connecticut, a town of 20,000 people and five elementary schools. In Ridgebury, Peter and Rebecca Coffey, along with other members of the Ridgebury PTA, organized a town-wide Mystery Festival!

In several detailed letters the Coffeys told us of their success. They wrote: "The event was organized and sponsored by parents, with the interest and support of third-, fourth-, and fifth-grade science teachers, the town science and curriculum coordinators, and representatives from other town PTAs." In addition to providing a number of excellent suggestions for adapting the activities, the Coffeys described their creation of an accompanying mystery video (and enclosed a copy). Their letter was euphoric:

The parent participation was called "magical" by one participant, and no one disagreed. This is an opportunity for parents and children to think together, play together, learn together, and work as equals on the solution to the crime. The ambiguity of the outcome reinforces this opportunity: no one can say, "See, I was right!" We also found that many more fathers got involved than with many PTA parent/child programs. Finally, by involving the parents who really enjoyed the event, the word about hands-on science gets spread far and wide quickly.

Several other large-group programs have been developed by the GEMS program, including Build It! Festival, a Microscopic Explorations event, an eco-mystery called Environmental Detectives, Math Around the World, and Bubble Festival. Any or all of them, along with FAMILY MATH or other programs (see page 130), make great vehicles for parent participation and the spread of active learning in science and mathematics.

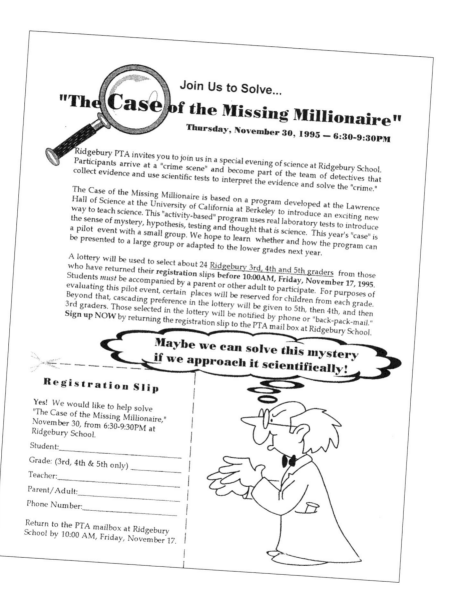

Join Us to Solve...

"The Case of the Missing Millionaire"

Thursday, November 30, 1995 — 6:30-9:30PM

Ridgebury PTA invites you to join us in a special evening of science at Ridgebury School. Participants arrive at a "crime scene" and become part of the team of detectives that collect evidence and use scientific tests to interpret the evidence and solve the "crime."

The Case of the Missing Millionaire is based on a program developed at the Lawrence Hall of Science at the University of California at Berkeley to introduce an exciting new way to teach science. This "activity-based" program uses real laboratory tests to introduce the sense of mystery, hypothesis, testing and thought that *is* science. This year's "case" is a pilot event with a small group. We hope to learn whether and how the program can be presented to a large group or adapted to the lower grades next year.

A lottery will be used to select about 24 Ridgebury 3rd, 4th and 5th graders from those who have returned their **registration** slips **before 10:00AM, Friday, November 17, 1995.** Students *must* be accompanied by a parent or other adult to participate. For purposes of evaluating this pilot event, certain places will be reserved for children from each grade. Beyond that, cascading preference in the lottery will be given to 5th, then 4th, and then 3rd graders. Those selected in the lottery will be notified by phone or "back-pack-mail." **Sign up NOW** by returning the registration slip to the PTA mail box at Ridgebury School.

Maybe we can solve this mystery if we approach it scientifically!

Registration Slip

Yes! We would like to help solve "The Case of the Missing Millionaire," November 30, from 6:30-9:30PM at Ridgebury School.

Student:_____

Grade: (3rd, 4th & 5th only) _____

Teacher:_____

Parent/Adult:_____

Phone Number:_____

Return to the PTA mailbox at Ridgebury School by 10:00 AM, Friday, November 17.

CHAPTER TWO

MATH AND SCIENCE EDUCATION TODAY
What's New Since We Were in School?

The Current Educational Context

It's a different world!

No need to tell you the world is very different from when we were in school—and it continues changing at an astonishingly rapid pace. Science research charges ahead, creating new knowledge faster than we can make textbooks that describe it. Thirty years ago, the world's pool of knowledge (the global "database," so to speak, of accrued discovery) doubled every fourteen years; today it doubles every seven years. (And thirty years from now, will it be doubling every year?...every hour?)

Not only is the speed of discovery increasing, the Internet and other modern forms of communication have hugely accelerated the rate at which knowledge is actually applied. Whole new content areas exist in medicine, genetics, microbiology, computer science, and technology, and these are playing an increasingly big role in our lifestyles, communities, and employment possibilities. Members of our society have to make a growing number of decisions as voters, consumers, and health-conscious individuals, about things like genetic engineering, the use of bovine growth hormone (BGH), over-fished marine species, safety issues related to local industries, fitness, pesticide use, and numerous risk assessments we make each day of our lives. These decisions all require a level of science knowledge and critical thinking. What's important for today's students to know? How can today's students best be prepared for their future?

The education system in the United States was crafted in the 1920s, when assembly-line and other repetitive-work jobs were the norm. Schools focused on teaching facts and basic skills. Only the top of the class was prepared for college and other, higher-level, "thinking-required" futures. Our current workforce needs are very different. More and more jobs require that people be able to learn, reason, think creatively, make decisions, and solve problems. Workers need advanced skills and the capacity for complex thinking. Factory workers are taking a greater role in collaborative, multilevel work groups, analyzing what they're doing and figuring out better ways of doing it. High-tech employers say they need staff who can learn new skills and knowledge on the job and by the week. Computers, needless to say, are everywhere. The Internet seems about to consume us. There seems no end to technological transformations in work and life.

Our education system hasn't grown at anywhere *near* the same pace as the rest of North American society and world technology—or in anywhere close to the same directions. This gap existed to some extent when today's parents were in school; the current gap is enormous. In addition, under today's social conditions schools face different challenges from those of our childhood. In our school days we might have been escorted to the principal's office for chewing gum; these days, hall monitors may include a sniffer dog and the principal may be dealing with stashed weapons. While proposed solutions differ, everyone agrees that we face extremely difficult challenges and urgently need to improve the current educational system. Our schools need to be far better aligned with the realities of today's world, and to make adjustments to better prepare our students for the future.

Budging the boulder, changing the system

The thing about societal systems is that they're big and they take a long time to change. Knowing the shortcomings is a far cry from implementing broad changes with a nation of children. Changing an educational system means changing things as fundamental and broad as what you see in the box below.

> ### THE FUNDAMENTAL COMPONENTS OF EDUCATION
> 1. How teachers teach (what's known as instruction)
> 2. How teachers are prepared for their profession (teacher preparation)

3. How teachers are provided with ongoing learning opportunities (what's currently referred to as teacher professional development)

4. What's taught in the schools (the curriculum)

5. The teaching tools that are provided for teachers (things like textbooks and equipment, known as the instructional materials)

6. The ways we determine what students are learning (assessment)

And of course changes as fundamental as these require a reassessment of how much money is needed, what it's needed for, and how teachers', administrators', and students' time can best be spent. There are so many interconnected parts in the educational structure, and so many components of the "school systems" of states, counties, and cities, that what's needed now is systemic change. The entire, complex enterprise to improve education overall in our nation is what's referred to as **educational reform**.

The pendulum of science and mathematics educational reform in the United States

In 1983, a highly publicized report by the National Commission on Excellence in Education, entitled *A Nation at Risk: The Imperative for Educational Reform*, found that "the educational foundations of our society are presently being eroded by a rising tide of mediocrity that threatens our very future as a Nation and as a people." The report shocked the nation by concluding that "if an unfriendly foreign power had attempted to impose on America the mediocre educational performance that exists today, we might well have viewed it as an act of war. As it stands, we have allowed this to happen to ourselves." National reaction to *A Nation at Risk* launched educators, politicians, the business community, and the public at large into what has since become a steady stream of reports, analyses, critiques, recommendations, concerns, complaints, and ideas—all focused on the need for educational reform.

More recently, in September 2000, the National Commission on Mathematics and Science Teaching for the 21st Century published another report to the nation, a cogent summary of key issues in science and mathematics education aptly entitled *Before It's Too Late*. The commission was led by famed astronaut and United States Senator John Glenn, and included outstanding educators, scientists, teachers, and political leaders. Their report begins, "First, at the daybreak of this new century and millennium, the commission is convinced that the future well-being of our nation and people depends not just on how we educate our children generally, but on how well we educate them in math and science specifically."

FROM *BEFORE IT'S TOO LATE: A REPORT TO THE NATION FROM THE NATIONAL COMMISSION ON MATHEMATICS AND SCIENCE TEACHING*

Parents:
- Do the mathematics and science achievement levels at your child's school compare favorably to achievement levels in neighboring schools?
- Do your child's teachers have the necessary background to teach the courses to which they are assigned?
- How does the teacher salary schedule in your school district compare to that of neighboring districts?

If your answers to any of these questions are at all disquieting, the following checklist can inform efforts to promote higher student achievement:

- Learn about the mathematics and science standards that children are required to meet in your state and get a clear picture of how well your child's school is doing in meeting these standards.

- Support the principal's efforts at your child's school to hire well-qualified teachers and to provide them with opportunities to continually improve their skills.

- Support increased funding for programs that support quality mathematics and science teaching.

In the years between *A Nation at Risk* and *Before It's Too Late*, the pendulum of education reform swung back and forth several times. We lived through "back to basics," "reading wars," "math wars," and more. While elements of these debates are not new, over the past two decades the nation has made a more sustained priority of the need for educational reform, elevating the debates to a new level. Unfortunately the polarization between educational reformists has often intensified too. The tug-of-war between these extremes in the United States has resulted in a flip-flop policy that's impeded long-term progress in educational reform.

The long-standing debate between educational "traditionalists" and "progressives" has run hot in many fields. In the Fall 2001 issue of *Education Next*, Lauren Resnick, a leading educator, summarized it this way:

> *Traditionalists* believe that there is a core body of knowledge that all students ought to learn: mathematical and scientific concepts, historical facts and interpretations, books that are part of our

shared American heritage. Traditionalists call for building the curriculum around such concepts and shared texts. *Progressives* are less interested in the specific knowledge or the texts to be studied; they are more interested in students' ability to use that knowledge and to find new information when necessary. Progressives often argue that there is too much information to imagine that anyone can master all of it, and they are—by and large—less concerned that everyone share a common core of knowledge.

Of course, as in any polarization, there are countless viewpoints in between. Many reformists classified in the progressive camp would strongly agree that there are key concepts, books, and facts that all children should learn. Conversely, many traditionalists acknowledge the importance of problem solving and independent thinking.

There's more to educational reform than the scrapping over education basics. The reform movement is playing out against the backdrop of a strongly autonomous national character, in which states' rights, individualism, and decentralization have played important roles throughout our history. It's likely that our country's version of educational reform will continue to be a political tussle on the mat, with a rainbow of perspectives, a diversity of vision, and no one set of solutions.

Within the whole picture of educational reform, separate stories play out for each discipline. For instance, the mathematics community has long debated the two ends of the teaching continuum: problem solving and basic math skills. The appropriateness of using calculators or not, or teaching long division or not, and the relative importance of rote memorization of math facts; these are all controversial in one circle or another. And with the nationwide focus on reading and literacy, the priority placed on teaching mathematics has plummeted. (This is even more true in science education.) Resources and time are being focused on subjects that will be "covered" in the standardized tests. (When you hear the phrase "teaching to the test," this is what it's about.)

Standardized tests: General achievement tests designed to measure how well a student has gained basic knowledge and skills taught in school, such as in math and science. You may recognize some of these; they include the Iowa Test of Basic Skills (ITBS), the Comprehensive Tests of Basic Skills (CTBS), and the Stanford Achievement Test Series (SAT-8, SAT-9, etc., depending on where it is in the series).

The "pit" in the pendulum; research at the core

Because the contexts of learning and schools are diverse, and because there's more than one right way to do things, educational reform isn't a straightforward problem. It's clear that much more research is needed. Unfortunately, much of the cry for one extreme or another seems to have been driven by belief systems rather than research-based results. Education is gut-important to each of us; to our children and our future, as individuals and as a nation. The debates, therefore, have been passionate and driven by more (or rather, less) than actual data. The results have been battles over strongly held beliefs rather than logical discussion of research-based results.

No matter your perspective, the gains and nose-dives of progress have led to a "snakes and ladders" history, as the pendulum of educational reform swings back and forth at least once a generation, if not more. That situation will continue as long as educational policy and practices are negotiated by politicians and based on beliefs and opinion rather than research.

There's enough in place for a good education!

The good news is that the system doesn't have to be totally transformed or perfect in order for your child to get a good—even a great—education. Your focus as a parent should be on the individual school your child attends, her teachers, the experiences you can provide for your child to augment her schooling, and the ways you interact with your child outside of school.

That said, it's important to know enough about today's important issues in education to help you successfully negotiate the current educational system with and for your child. This section goes on to describe some of these key issues (in many cases substantially different from those of our school days) and what research results do tell us.

Today's Important Issues in Education

Achieving balance in math and science education

In some communities, activity-based approaches in mathematics and science education have been criticized by those who believe in traditional, lecture-based teaching methods. Traditionalists argue that teaching basic skills, *not* innovative, hands-on, inquiry-based approaches, should be at the core of math and science programs ("What was needed in the past is still needed today").

Progressives contend that rote learning, with its sole emphasis on factual knowledge and skill building, doesn't develop conceptual understanding or problem-solving skills ("What failed to work in the past for many students will still fail to work today"). Progressive educators certainly agree that students need to develop and learn basic skills; but they also see the need for students to develop critical-thinking abilities, in order to apply their skills to non-routine problems.

The debate heats up when test scores fall below par. If a progressive curriculum is in place when the scores go down, traditionalists point to the low scores and call for a "back to basics" program. If a traditional program is in place, progressives point to a culprit traditionalist policy (the tendency for students to memorize uncomprehendingly) and call for programmatic changes. Each group is inclined to blame the other for poor performance—often before enough time has elapsed to truly evaluate what students have learned! This further polarizes the two groups and throws gasoline on the debate.

Neither an exclusively traditionalist nor a solely progressive approach provides the balance that strong mathematics and science programs need. The common ground would be to create math and science programs that include a balance of skills, concepts, and problem solving.

A growing trust in research

Teaching has been considered an art for many years, something individuals learn to do well in their own ways. While that will always be partly true, and there's much to learn from excellent practitioners, it's now widely accepted that there's more to effective teaching than raw talent. We need to look to research to understand what makes for good teaching and effective learning. In particular, we need to learn how to narrow the achievement gap that exists in our country between students of different socioeconomic levels.

> **"Best practices"**: The most current research-based methods of instruction and assessment, intended to address how children learn best.

Thanks to an increased research focus, we now know more than ever about "best practices" in teaching—strategies, materials, and teaching approaches that promise the best learning results for students. There have been literally thousands of research studies focusing on teaching and learning, resulting in many notable improvements in educational practices in the last generation. But while we know more than we ever have, there's still a lot to learn. Many in the field of educational research believe we're on the edge of a knowledge breakthrough that will transform the field yet again.

What research says about how students learn best

For all the controversy in the field of education, there's considerable agreement about how students—and all people—learn best. Here's our synthesis of what the research shows and our own experience over many years of designing programs for learners of all ages and abilities:

HOW PEOPLE LEARN BEST

- **"Active" learning is most effective.** When a child interacts directly with materials and ideas, rather than just hearing or reading about them, he understands better and retains longer. Note that "active" can mean "hands-on," in which children interact with materials, but it can also mean "interactive," in which they interact with *ideas*, through discussion, writing, and reflecting. An effective teacher involves students in learning-by-doing, working in groups, discussing, writing, and reflecting—in *addition* to reading and listening.

- **Understanding goes beyond vocabulary.** Facts and formulas are important in mathematics and science—but memorizing vocabulary or math tables doesn't explain or make sense of the concepts behind them. Without this deeper understanding, knowledge is easily forgotten. True understanding involves a much deeper approach to learning about concepts, and this takes more time. Effective teachers present topics in greater depth in order to deepen student understanding.

- **There are recognized phases of learning.** There's a progression to learning; this should be allowed to unfold in its natural sequence. A learner needs a chance to explore ideas, receive information, and then apply what she's learned to new situations. She needs to come up with her own questions and investigate for herself. Effective teachers know how to sequence these types of activities to maximize student learning.

- **Learners of different ages have different needs.** Young children need lots of free exploration and time to learn through experience. Older children are capable of more abstract thinking, and can apply what they've learned to more complex situations. But research shows that all learners, not just those who are young, need to interact directly with materials. Effective teachers provide students of different ages with the experiences they need.

- **A variety of learning experiences is best.** Like good nutrition, best learning comes from a "balanced diet" of experiences—reading, discussion, hands-on experiences, listening, being active. Effective teachers provide a variety of learning opportunities. A long-range perspective about your child's education, understanding that her education will be a lifelong "running sum" of all her experiences, helps you maintain

> calm when a teacher is providing only certain kinds of experiences. It can also guide how you choose to balance your child's school experiences with out-of-school opportunities.

These findings represent diverse points of view. As with so many things, of course, research findings may be perceived differently by people with particular agendas or belief systems. People decide what they want to see.

From our point of view, three things seem clear. First, anyone who claims to have all the answers, and dogmatically asserts that one educational approach is the "only way to fly," is, at best, mistaken—and at worst, has a hidden (and biased) agenda. Second, following from the first, a flexible, non-regimented approach to curriculum and instruction seems the most fitting embrace of the vast range of ways in which people learn. And third, some real humility is in order. The world is constantly learning more about how people learn. Researchers analyze tapes showing effective teachers leading constructive discussions; they interview students before and after they've mastered a complex concept; they conduct breakthrough scientific research on the human brain itself.... There is, happily, no end to the ways in which we expand our understanding of what makes us tick—and learn.

Testing: reasons and value over time

Historically, testing was used as a way of identifying "the best and brightest" students, those who would move into a limited number of slots in secondary and post-secondary education. In an era when educational resources were very limited, this may have made sense. But today the goals are to increase, rather than limit, the number of students who succeed and go on to higher education. With this change in context come different needs and purposes for testing. Learning what students understand (or don't!) provides important feedback on instruction (*how* a teacher teaches) and curriculum (*what* a teacher teaches). This feedback is important for the teacher, the student, the parent, the school, the school district, and the state.

One of the current practices in education (new since today's parents were children) is to provide a *variety* of tests ("multiple measures," they're called), so that we can assess a broader range of student knowledge and ability. Multiple-choice tests represent a traditional form of assessment. Tests that involve students in real-world tasks are currently referred to as **performance tasks**.

Say a person goes to the Department of Motor Vehicles to get a driver's license. There he encounters two kinds of tests. One is a written multiple-choice exam, which tests for knowledge of driving laws. The other is a true driving test, which assesses a person's ability to operate a vehicle using his knowledge of those laws and his good judgment. Which test is most valuable? It's hard to imagine not testing in both ways, to assess both kinds of knowledge! We want to know that students are developing both the basic knowledge of driving laws (What's the speed limit on this road?) and the critical-thinking skills needed to drive safely (What if that truck pulls out as I pass?).

All kinds of learning are important, from mastery of factual information to critical-thinking skills and the understanding of complex concepts. It's important to find out, for instance, that children know their math facts (like speed limits); it's equally important to learn if children know how to think critically and solve complex problems (like driving itself).

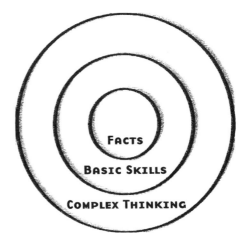

How testing has changed

Educators currently use a variety of methods to assess what students know. Some of these resemble traditional tests; others aren't always recognizable as such. All of them are broadly referred to as **assessment**. Multiple-choice and short-answer questions are quite adequate for assessing knowledge of facts and certain kinds of conceptual knowledge. Performance tasks are better for assessing the kind of complex reasoning that demonstrates that a student truly understands. The kind of assessment used depends on what kind of information is desired.

Assessment is now commonly woven into all stages of education: **before** new material is presented (to assess existing knowledge), **during** the learning process (to monitor how well a child is grasping a concept), and **after** the lesson's been taught (to check whether the child understands the new material). This lets a teacher assess and modify the learning experience in time to make a difference—rather than discover a problem just as it's time to move on to another subject.

Parents sometimes wonder "where the tests are" nowadays, thinking that there's not as much testing as there used to be. Because today's teachers use multiple measures to inform themselves about student progress, it's not always clear what the test is or when it's occurring. Assessment can take the form of written tests, portfolios of work (just as an architect or artist might have), experiments, projects, or teacher observations. Assessment now reflects the number of ways in which a student's skills or subject knowledge can be observed and/or measured. Assessment can be formal, in which case a student knows it's a test, or it can be informal, in which case it's indistinguishable from the curriculum (this is called **embedded assessment**). Current "tests" may not always look like tests, but that doesn't mean they're not equally rigorous.

An effective teacher is constantly assessing students' progress through daily and weekly home and class assignments as well as ongoing observations. The approach of your child's teacher to assessment is an excellent subject to raise and discuss as the school year gets under way.

Assessment: Testing, but more varied and ongoing than the big, traditional, end-of-course "finals."

a) **Alternative assessment**: Any means other than traditional tests to assess what a student knows and can do. Includes performance-based assessments (below), portfolios of student work, and other methods.

b) **Performance-based assessment** (also called "authentic assessment"): Assessment that requires a student to perform a task—write an essay, create a graph, conduct an experiment, etc. This is increasingly replacing multiple-choice tests. It provides a more realistic and "whole-picture" measure of how a child is learning.

"Grading on the curve" and more: current methods of scoring work

Traditionally, tests and other kinds of student work were graded relative to other students' work. This meant that tests were scored and then assigned grades (or percentile ranks) according to a *group* of test scores. So, for instance, the top ten percent of students in the group might score an A—no matter what their actual scores were. Or they'd be assigned a percentile rank, such as ninetieth percentile, meaning that they'd scored in the top ten percent of students in the class (or other comparison group). This method of scoring tests, comparing a student's work against an average of students in a larger group, is called **norm-referenced scoring.** Many tests and scoring procedures still use this approach.

> **Norm-referenced scoring**: A method of scoring tests by comparing a student's work against a national average of students at the same grade level.

> **Criterion-referenced scoring:** A method of scoring tests by comparing students' work with established standards or criteria described in a rubric (scoring guide).

The advantage of norm-referenced scoring is that a student knows how her work compares to other students' work. One disadvantage is that it doesn't provide information about a student's level of mastery in a subject, or even necessarily reflect a school's academic quality. Take the case, for instance, in which states offer incentives for schools that improve their ranking using a norm-referenced scoring system. **Even if every school in the state dramatically increases its test scores, fifty percent of the schools will still score in the bottom half.** This is a serious limitation of norm-referenced scoring.

In the last decade or so, educators have begun using **criterion-referenced scoring**, a method that compares a student's work with established standards or criteria. These criteria are described in scoring guides, sometimes called **rubrics.** (See Appendix C, pages 135–137, for examples of rubrics developed for programs at the Lawrence Hall of Science.)

The advantage of criterion-referenced scoring is that the focus is on achieving high standards of work, rather than competing with other students. (It makes sense that any student should be able to achieve an A or other top score if her work meets a certain quality standard.) The disadvantage is that students (and parents) can't know how a student's work compares with the work of other students.

Increasingly, teachers are sharing rubrics (scoring guides) with their classes so students can understand the definition of quality work and how their work will be judged. A huge benefit is that students begin internalizing those standards and honing their ability to critique their own work. These are essential life skills that serve children long after they've left school.

> **Rubric:** A scoring guide for a test or other assessment task.

What's meant by educational "standards"?

Definition and history

Educational standards are created to define a vision of education. There are standards for teaching, for professional development of teachers, for assessment, and for program development. The most common standards—what's generally meant when people discuss "standards" in the news or in educational articles—are **content standards**. These set out to define what students need to know at different grade levels.

The attempt to agree on standards has been a highly political process in our country. A few other industrialized countries manage to set national standards relatively easily, without much controversy (or input, for that matter—

which may explain that). Given the strong culture of autonomy and democracy in this country, however, the "standards battle" will continue to be a hard and complex one.

There are many different sets of standards in the United States. The most commonly applied standards in mathematics are the *Principles and Standards for School Mathematics*, created by the National Council of Teachers of Mathematics (NCTM). In science, there are the *National Science Education Standards*, created by the National Academy of Science's National Research Council (NRC), and the *Benchmarks for Science Literacy*, created by the American Association for the Advancement of Science (AAAS) Project 2061. Every state but one (Iowa) also has its own standards in every major subject. Many school districts have *their* own standards. State and district standards are usually fairly derivative of the national standards mentioned above, but they sometimes differ substantially.

> **Standards:** National- and state-determined goals for what students should know, understand, and be able to do in various subjects at different grade levels.

In Appendix A (pages 123–129), we've provided samples from the following national standards:

National Science Education Standards: www.nap.edu/books/0309053269/html

Benchmarks for Science Literacy: www.project2061.org/tools/benchol/bolframe.htm

Principles and Standards for School Mathematics: http://standards.nctm.org

You can look up your state's standards on the Internet (the clearest, most intuitive site is Education World's, at www.education-world.com/standards/) or in the reference section of most libraries.

The explosion of standards-based reform

The standards movement in our country started in the late 1980s and gave rise to a sweeping approach called **standards-based reform**, or **standards-based education**. In this approach, the knowledge students are expected to acquire is defined in a set of standards. Teachers present a "standards-based curriculum" that enables students to acquire that knowledge. And there are standardized tests, aligned to the standards, that test for that knowledge.

To the surprise of many educators, standards-based education has practically become national policy—not because standards are mandatory, but because most federal funding (as well as many other sources of reward and funds) is tied to standards-based

> **Benchmarks:** Standardized kinds and levels of understanding schooling should produce in each subject, at each grade level. Benchmarks determine both the content and the level of understanding students are expected to master as they move through the K–12 system.

57

programs and standards-based achievement.

> When a school's funding depends on how its students perform on standardized tests, or when a student can only advance to the next grade once he's demonstrated knowledge of the standards, it's called "**high stakes**" standards.

In theory, standards-based education, with its emphasis on a defined set of standards, makes a lot of sense and seems straightforward enough. When it was first proposed, it appealed to both sides of the political spectrum as a way to promote higher expectations for *all* students (and a better-educated workforce). It appeared to focus—and, some believed, actually raise—the level of content being taught.

However.

Standards-based reform; not so simple

Standards-based education operates on these three principles:

1. That there are agreed-upon standards of what all students should know.
2. That students are presented a curriculum that teaches what's in the standards.
3. That there are standardized tests, aligned with the standards, that test for that knowledge.

Simple? Seems like it should be. But consider these questions:

- **Which standards should be used?** National, state, or district? They're often very different.

- **What mix of knowledge and reasoning skills should standards be based on?** Should standards be based mostly on things that need to be memorized or things that require reasoning? Currently, some sets of standards emphasize a huge number of disconnected facts and skills. Critics believe this doesn't live up to the vision of what standards were supposed to create—an intellectually demanding level of knowledge; the bow wave of the educational reform effort.

- **Which knowledge?** Should students learn about state capitals? biotechnology? the periodic table? the orbit of the Moon? why earthquakes happen? No teacher can do it all, and hard decisions need to be made about what there's *not* time to teach. National standards, especially in science, have tended to be voluminous (in part because they were put together by a broad range of scientists and mathematicians, none of whom wanted their particular discipline left out). It would take a teacher several lifetimes to teach them all! So even schools and school districts that can agree to embrace the national standards must decide what subset of those standards to include.

- **Who chooses the standards?** The matter of who gets to decide what standards are used is a highly political question. Deborah Meier, Senior Annenberg Fellow at Brown University and author of *Will Standards Save Public Education?* (Beacon Press, 2000), believes standards should only be set by a group of adults who know the particular students for whom those standards will be used: the teachers in a school, the students' parents, or perhaps an advisory group. Educator and advocate Susan Ohanian, author of *One Size Fits Few: The Folly of Educational Standards* (Heinemann, 1999), argues even more vigorously than Meier that individual children don't fit into standard categories, and that only a child's individual teacher can know what's right for that child. Critics argue that this kind of local control introduces a potential for "wrong decisions" resulting in low-quality and wildly diverse standards—and therefore invalidates standardized tests as a measure of what learning has occurred.

- **For which students?** Should standards be written so they're attainable by all students, creating a "floor" of knowledge and abilities, or should they be challenging enough that they're attainable only by the best students, creating a "ceiling"? Is this a matter of graduating the best-prepared young people, or a form of elitism?

- **What role should standards play?** Should a school just be aware of standards, and try to work them into its curriculum, or should it be *required* to meet them? And who's accountable for meeting standards—the schools? the teachers? the students themselves?

- **Do all students have an equal opportunity to learn?** Because the whole process of standards-based reform has moved so fast, few teachers and schools have been able to create a curriculum that's well aligned to the standards. There hasn't been the time, professional development, or funding to support that monumental change. As a result, a key element in standards-based reform—that students will have the opportunity to acquire the knowledge in the standards—has been spotty at best. Many believe that failure to invest the time and resources in teacher professional development and curriculum planning has created a fatal flaw in the process. Students are not getting equal opportunities to learn.

- **Are there good standardized tests aligned to standards?**
Lauren Resnick, in her article "The Mismeasure of Learning" (*Education Next*, Fall 2001), flatly states: "No amount of getting standards right will make much difference when states and districts are calling for teachers to raise scores on tests that do not match the standards anyway." In practice, few states use tests and other criteria that are well aligned to the standards. In a study conducted between 1999 and 2001 by Achieve, an organization of state governors and business leaders, it was found that **not a single state had tests well aligned to the standards in mathematics and literacy**. In *The Schools Our Children Deserve: Moving Beyond Traditional Classrooms and "Tougher Standards"* (Houghton Mifflin, 1999), author Alfie Kohn points out how many of today's standardized tests stress disconnected knowledge and tasks that require low-level thinking. As teachers scurry to prepare their students for these tests, the curriculum gets absurdly "dumbed down." Author Peter Sacks, in *Standardized Minds: The High Price of America's Testing Culture and What We Can Do to Change It* (Perseus Books, 2001), also makes the case against standardized tests that perpetuate a focus on "mile-wide, inch-deep" curriculum rather than promoting teaching for solid knowledge of important topics.

Depending on the quality of the standards that are chosen, standards-based reform could be an enlightened path or a regressive one. And as long as the standardized tests are driving this accountability system, and are misaligned to standards or represent otherwise shallow measures of learning, standards-based reform will not occur effectively or as designed.

It's anything but simple.

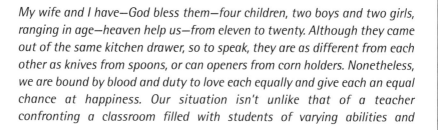

In a Family Way
by Howard Good
Journalism Program Coordinator, State University of New York at New Paltz

My wife and I have—God bless them—four children, two boys and two girls, ranging in age—heaven help us—from eleven to twenty. Although they came out of the same kitchen drawer, so to speak, they are as different from each other as knives from spoons, or can openers from corn holders. Nonetheless, we are bound by blood and duty to love each equally and give each an equal chance at happiness. Our situation isn't unlike that of a teacher confronting a classroom filled with students of varying abilities and interests. He or she is expected to instruct them all, the underachiever as well as the overachiever, the dim as well as the bright.

But how? My wife, a wise woman with an ironic sense of humor, believes the key to raising children, especially teenagers, is to figure out what it is they want to do and then, if it's legal and within the credit card limit, advise them to do it. She claims this greatly reduces the risk of parental stroke. And she may be right. Certainly when we tried to bend our oldest to our will, the results were less than impressive. He still got speeding tickets, still failed courses, still hung out with kids who looked like extras from a "Mad Max" movie. We finally realized that we couldn't order him to grow up—or, rather, we could, but only at the cost of further eroding our already crumbling relationship with him. Yes, we are disappointed that he flunked out of one college and dropped out of another; yes, we are frustrated that he simply shrugs whenever we ask about his long-term plans. But we are also increasingly aware that life unfolds without much regard for parents' dreams and wishes. He will grow up, just on his own timetable, not ours. If there is a lesson here for educators, it is perhaps that you can't force children to become who they aren't ready to be. The current nationwide trend toward more frequent standardized testing seems to ignore this basic fact. Its supporters envision education as a kind of automated assembly line, all students in the same grade learning the same things in the same way with the same outcome. They take little account of the child who develops along a different path or at a different pace.

To go back to my oldest, I have often felt that he was at a disadvantage throughout his years of public school because his best qualities weren't among those generally measured by tests or prized by teachers. He has, for example, an uncanny ability to relate to animals, big or small, wild or domestic, fur-bearing or scaly. I have seen him rescue panic-stricken birds that found themselves trapped in our screened-in porch, calming their frightened flapping with his own lack of fear and then scooping them up in his bare hands and setting them free. The birds recognize what standardized tests can't: his gentleness and sympathy. A district may have high test scores, but how good, really, is a school system that treats as insignificant the very qualities a cruel and bloody world most needs?

My second son, unlike the first, is an outstanding student by all the conventional measures: test scores, grade point average, class rank. I would have to add together my grades in two high school courses to equal his grade in one. The real irony, though, is that his school career began rather inauspiciously. Following kindergarten screening, he was assigned to a class for at-risk students. When my wife and I questioned the placement—and who knew him better than we did?—the principal said he could go into a regular kindergarten class, but if he struggled, it would be our fault. Intimidated by the prospect, we backed down.

That this story has a happy ending, with my son eventually casting off the inaccurate label slapped on him, doesn't mean it is a happy story. To me, it illustrates the difficulty too many parents have in making themselves heard above the groaning of a constipated educational bureaucracy. These are challenging times in which to raise children, and parents should at least be able to count on schools as their allies.

I never feel more like I need an ally than when I look at my oldest daughter, soon to be 15, parading around with her bellybutton fashionably exposed. Sometimes it seems the whole culture is conspiring to turn her into a mall-going, instant-messaging ding-dong who derives her role models from MTV, her insights from movies, and her greatest gratification from shopping. And where can a parent get help in beating back this sea of cultural crud? Not necessarily from schools. The ideology of relentless, ever-escalating consumption—what novelist Ted Rall bluntly called Americans' obsession with "earning as much as they can so they can buy as much worthless crap as they can as quickly as they can"—has penetrated even there. School leaders now often promote education to a skeptical public by emphasizing its role not in preserving democracy or building character, but in increasing the purchasing power of graduates.

Isn't it more important to be a good citizen than a smart shopper? Shouldn't schools be a haven from, and not simply an extension of, the prevailing commercial culture, with its over-the-top worship of material success? When so much of society has become infected with the virus of greed, shouldn't teachers, entrusted with the mental and moral development of the community's children, have an obligation to carry forward the true values of life?

For the sake of my youngest child, I hope so. She is just entering middle school, which, to the overactive imaginations of anxious parents, seems like the dark woods in a Grimms' fairy tale, inhabited by evil spirits, hairy beasts, and plotting witches. Research shows that the American middle school is, in fact, a perennial trouble spot; the casualties are self-worth and academic achievement.

I do not have an easy solution—I leave that to the so-called experts—but I do have a suggestion, and it won't raise taxes a single cent or require state education department approval. It is that school boards, administrators, faculty, staff, parents, and students acknowledge that our fates are inextricably intertwined and treat each other accordingly, as members of one big, raucous family. Of course, we would still have problems and conflicts. What family doesn't? But we would also have the encouragement of knowing that we are, at bottom, more alike than different; that we are bound to each other by silver cords of memory and tradition; that indoors or out, we are surrounded by people who would lift us if we fell and cheer us if we succeeded and love us no matter what.

This article is reprinted with permission from *Teacher Magazine*, November/December 2001.

Aligning instruction with testing

Over the past decade the issue of standards for schools has become as much a political issue as a technical one. One of the unfortunate consequences is that much of educators' attention is now fastened on improving test scores, with too little attention being paid to how to ensure that students learn more. This is akin to business leaders worrying about how to get profits up without worrying about the quality of the products they produce or the customers they serve. It also has turned debates about standards into debates over which tests to use and which test is best....Concern with measures of learning is so great that in some states it is difficult to get any serious consideration of ways to improve student experiences precisely because teachers and administrators feel an urgent need to improve test scores regardless of what students actually learn.
—*Phillip C. Schlechty,* Shaking Up the Schoolhouse: How to Support and Sustain Educational Innovation *(Jossey-Bass; 2000)*

Active, in-depth learning and high-quality standards work well together—but only if the testing and other student assessments are designed to provide solid information on student knowledge and abilities, not just on mastery of technical vocabulary or other rote learning.

In recent decades, researchers have learned a lot about "authentic," or "performance-based," assessment. They've also made strides in learning how to lessen cultural and class biases, making testing more equitable. It's a challenge to incorporate these ideas into tests administered to millions of students, but it's the only way to ensure that standards, curriculum, instruction, and testing are equitable, meaningful, and aligned with each other. Only when this is the case will we have an accurate gauge of student progress. And only then will we know whether or not students are reaching the high level of scientific and mathematical literacy demanded by our times and the challenges of the future.

Let's revisit the requirements for a driver's license. The written test on rules and regulations is important and needed. It helps the potential driver learn the information, and provides some evidence to the DMV that this knowledge has been studied and acquired. But you don't teach driving with paper and pencil—actual practice is needed. An examiner can't determine if someone is a competent, qualified driver without seeing that person drive. A state's "standards" for a driver's license need to be of sufficiently high quality to protect everyone on the road (and all pedestrians!); the tests need to provide at least some assurance that the person being tested can meet those standards. In this same way, multiple measures are needed to accurately assess a student's (or a school's) overall academic performance.

Beyond the pendulum

What is measured in high-stakes assessments has a profound effect on human behavior. We must be exceedingly careful to make sure we measure what counts. We cannot expect major improvements in education without major changes in our assessments.
—Dr. Bruce Alberts, President, National Academy of Sciences

As the pendulum of education swings back and forth, some of its motions are nourished by new research and actual experience in schools with students and teachers. But there are other, less positive influences—like political agendas or axes to grind. And the sensationalism that's part of mass-media culture in the United States also has its echoes in education. A new catch-phrase or heavily promoted technology can sometimes become a proverbial bandwagon, gathering followers who lose sight of the larger picture as they jump on a technological advance or a "new" way to teach reading that promises to solve all the problems.

The huge impact of new technologies will continue to influence education in many ways, but experience has taught us that many of the initial claims about computer learning were, at best, inflated. The task of a conscious citizen, of an involved parent, is to be aware of new developments and urge schools to take advantage of their strengths, but not to overestimate their ability to solve all problems. There's no panacea or simple answer for the complex human enterprise of education. The art of teaching—the nurturing of wisdom—takes time, energy, experience, patience, and persistence. An artful weaving or marriage of approaches once considered to be mutually exclusive (phonics and whole language in reading, for example) often turns out to be not only a more balanced vision, but more educationally effective than either approach by itself! We must look beyond the pendulum to envision a positive future.

Making Sense of It All

Learn what's important

Understanding what to value in educational reform puts you at ease and gives you real power. If you consciously choose to read or listen to items about reform when the subject comes up in the news, rather than tune it out, you'll be pleasantly surprised to find it's not mysterious. You'll instinctively form opinions—you're a parent, after all! You can decide what matters most to you about educational reform and stay in touch with those things.

But be careful of the news, too—especially on the subject of test scores and comparisons. Bring a critical sense to what you read or hear.

Not all "news" is what it seems.

QUESTIONS TO ASK YOURSELF ABOUT TEST-SCORE REPORTS

- What was the test designed to measure—factual recall or complex thinking?

- Were students being tested on what they'd been taught?

- Was there a connection between what was tested and what the state or national standards say is important?

- Were students tested in a language they understood well enough?

- Was the test measuring rapid recall or thoughtful response?

- What relevance did what was tested have to desired workforce skills?

- Were the students practiced in test-taking strategies?

Look behind the headlines

There's been a widely publicized decline in average SAT (Scholastic Aptitude Test) scores among college-bound students since 1975. This is factually true...but also a distortion of a more complex situation. For one thing, more people than ever before in the United States hope to go to college—so, many more are *taking* the SATs. Of this increasing number of test takers, a greater percentage rank in the bottom half of their high school classes. In 1993, Sandia National Laboratories conducted a study comparing 1975 SAT scores to 1990 scores, only looking at results from students from the same class rank and gender. It found that when these factors were controlled, the average SAT score for this "traditional test-taking population" had actually INCREASED by thirty points.

This is just one example of how truth about education—particularly standardized-test results—can get distorted in the news. Here are some pointers for bringing a critical sense to what you read or hear about this topic:

USING YOUR JUDGMENT TO ASSESS NEWS REPORTS

BEWARE "SCARY" STATISTICS.

The media feeds on sensation, and shocking school statistics make good stories. It's not uncommon for reports to use very generalized information—input from many different student bodies for example, dumped into one statistic—and draw misleading conclusions.

In one documented example, an international exam tested North American students for their knowledge of algebra—a full year before U.S. schools teach it! Not surprisingly, those U.S. students "tested poorly" in algebra! In another instance, statewide standardized-test results seemed to imply an alarming lack of student knowledge...but nowhere in the reports was it mentioned that students had had twenty minutes to answer sixty questions. Rapid recall and real knowledge aren't the same.

WEIGH THE VALUE OF STANDARDIZED TESTING.

A well-designed test (focused on highly valued, well-taught knowledge and skills, and administered to minimize bias) can provide important information. But even then, the cause of a "bad" test result—or even the value of the result in predicting a student's future—is not at all clear. It's well established, for example, that a high score on the SAT is NOT the best predictor of future success in college. (In fact, some universities are dropping the SAT score from their admission criteria.) So it's important to remember that test scores are only one of many means of assessing a student's knowledge.

CONSIDER THE SOURCE.

A surprising number of organizations have a vested interest in slanting results to reflect their thinking. This doesn't make for neutral reporting, and the published "findings" can be misleading. Administrators, teachers' unions, and even state education depart-ments may have reason to slant the interpretation of test results.

Groups unhappy with current accepted approaches to education may be motivated to emphasize the findings of one test over another, findings that imply the state of education isn't good enough. Groups that feel education is vastly underfunded may do the same, in hopes more money will be allocated. Even groups satisfied with the state of education to date, but determined that standards remain high or even increase, may want to send a cautionary message. These and other groups naturally draw conclusions that support their perspectives.

There's been careful and credible research on the need for improvement in the state of education in this country, but it's important to seek out reports that avoid any one group's particular agenda or personal goal. Reform must be based on real needs suggested by many factors, not just test results. On the home front, if you have questions about a test your child took or what the statistics mean for your child's school, ask to speak to the principal. Maintain perspective, and stay informed.

The time to act

As the century gets rolling, it's our profound hope that our society will find its way past the worst, most destructive aspects of the recurring pendulum-like debates and toward a new synthesis of best practices in education. Unless this happens, it's unlikely that our educational system can bring the majority of our citizens to a level of sufficient scientific literacy to make crucial decisions.

Before It's Too Late, the Glenn report mentioned on page 47, urges that "the time to act is now." The report quotes an ancient Chinese proverb: "The best time to plant a tree is twenty years ago; the second best time is today." The report goes on to say:

> At the beginning of the 20th century, schools in our country underwent profound changes as they adapted teaching and curricula to the demands of industrialization and the needs of a growing population. At the beginning of the 21st century we look to an even greater challenge. Our schools, and the teachers who imbue them with life, must find ways to produce higher levels of proficiency in mathematics and science in all their students. The American people must do this because our world and our times require it. If we do not plant this tree today, we will surely have to do so tomorrow, next month, or next year—at a higher price.

Overall, the report takes an optimistic view, emphasizing that the time is right for change. It places very strong emphasis on improving the teaching of math and science by devoting much greater resources to the professional development of teachers, improving teaching conditions, and raising teacher salaries substantially. The past two decades of experience in educational reform have demonstrated that "tinkering around the edges" is not enough, the report states. And polls show that many Americans place the need for improvement in education high on the agenda.

Over the next decade schools will replace two-thirds of their teachers, as older teachers retire, and many college students are expressing an interest in teaching. Much has been learned in recent years about what works best in science and math curriculum, and *Before It's Too Late* points out that many promising approaches are already underway. "We now have the chance to bring this knowledge together in a mathematics and science education effort that will serve as a powerful tool for change. Ignored research is worthless; unused tools can only rust….As the Chinese proverb says, the time to plant the tree is now—*before it's too late*."

CHAPTER THREE

PROVEN STRATEGIES
Supporting Your Child's Success in Math and Science

Approaches That Work for Every Parent

Vive la différence!

Each child is a unique and complex blend of individual personality; so is every parent! Keeping this marvelous human diversity in mind will help you develop effective ways to help your children succeed in math, science, and all other educational realms.

The word "style" has both strengths and limitations. We're going to use it to talk about "parenting approaches" and "learning abilities," because the word "style" is in common educational use—and it does contain the positive idea that we can change and combine different styles (mix and match, so to speak!).

As complex human beings, all of us have elements of many different learning—and parenting—approaches. As we talk of different "parenting styles" and "learning styles," it's important to keep in mind that the names researchers give to such approaches aren't intended to describe any *one* person, or to judge one style always superior to another. The classifications you'll read below are designed by researchers to describe some important findings about differences in how learners learn and how parents parent. We share them here to encourage you to think about different parenting approaches and find out more about your child's special mix of learning styles.

Parenting styles—where do you fit in?

[Good parents] watch and guard: they remove stumbling-blocks: they manifest approbation and disapprobation: they express wishes, but, at the same time, study the wishes of their little people: they leave as much as possible to natural retribution: they impose no opinions, and quarrel with none.
—Sociologist Harriet Martineau, 1835; Montgomery, Alabama

Research tells us that some parenting styles are significantly more effective than others in supporting a child's success in school. Laurence Steinberg, in his eye-opening book *Beyond the Classroom* (Touchstone, 1997), describes persuasive findings about this, summarizing the work of a number of research teams over many years. Steinberg describes three main parenting styles, which he classifies as **authoritarian, permissive,** and **responsive.** Within these three styles, he identifies dimensions along three ranges:

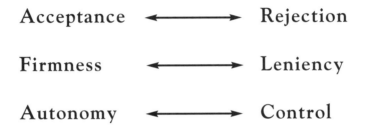

No parent falls strictly into one extreme or the other of any given range— we all fall somewhere in between and may differ on varying issues. But Steinberg does conclude that a tendency toward the **acceptance, firmness,** and **autonomy** end of these ranges has a STRONG correlation to academically successful children. Over the years, research has strongly tended to confirm that this combination of parenting dimensions leads to children who are successful in school. Here are some important findings about parenting tendencies—whether the parent's overall style is generally authoritarian, permissive, **or** responsive:

> ### PARENTING STYLES THAT SUPPORT A CHILD'S SUCCESS IN SCHOOL (AND LIFE!)
>
> - Parents who are **accepting** are responsive, involved, liberal with praise, and affectionate. (Rejecting parents, as you can imagine, typically act in opposite ways.)
>
> - Parents who are **firm** have high standards, set clear limits, determine consequences for behavior, and are consistent. (Lenient parents provide little or no structure for their child.)

- Parents who **promote autonomy** in their child show that they value self-expression and don't expect their child to have the same opinions they do in order to win affection. Parents who value autonomy emphasize independence over obedience. (Controlling parents demand unquestioning obedience from their child.)

Studies show that the greatest positive impact on a child's success occurs when all three of these dimensions occur together. The child of an accepting, firm, **and** autonomy-promoting parent is MOST LIKELY to succeed in school.

Your child's learning style

What is "intelligence," anyway?

In his book *Frames of Mind: The Theory of Multiple Intelligences* (Basic Books, 10th-anniversary ed. 1993), Harvard psychology professor Howard Gardner challenged the widely held notion that intelligence is a single general capacity, possessed in varying amounts by us all. Instead, building on the findings of modern brain science and psychology, he proposed that all human beings possess at least **seven** "intelligences." It was revolutionary.

Gardner sees intelligence as a biological and psychological potential that's realized differently by different people, according to each person's experience, education, social environment, and other factors. Viewed in this way, human intelligence is not restricted to the narrow language and math abilities measured by most standardized tests—what we traditionally call "school smarts." As Gardner's been quoted as saying, "It's not how smart you are, but *how* you are smart." Not surprisingly, educators have shown great interest in the idea of multiple intelligences, and the theory has survived and evolved considerably since Gardner first published it in 1983.

Here are the seven intelligences described by Gardner, briefly summarized:

SEVEN KINDS OF INTELLIGENCE TO LOOK FOR IN YOUR CHILD

- **Linguistic Intelligence:** The capacity to use words effectively, whether orally or in writing. This includes the ability to manipulate the structure and syntax of language, the sounds of language, the meanings of language, and the practical uses of language, such as for explaining, remembering, persuading, etc.

- **Logical-Mathematical Intelligence:** The capacity to use numbers effectively and to reason well. This includes awareness of logical patterns and relationships, functions, and cause and effect.

- **Spatial Intelligence:** The ability to perceive the visual and spatial world accurately, including sensitivity to color, line, shape, form, space, and the relationships between them. Includes the capacity to visualize, make graphic representations, and orient oneself in spatial surroundings.

- **Bodily-Kinesthetic Intelligence:** The ability to use one's whole body to express ideas and feelings, and the ability to fashion or transform with one's hands. Includes coordination, balance, dexterity, strength, flexibility, speed, and other physical skills.

- **Musical Intelligence:** The ability to perceive, distinguish between, and express oneself in musical forms. Includes sensitivities to rhythm, pitch or melody, timbre, and tone color. Can apply to either an intuitive grasp of music or an analytic, technical understanding of it—or both.

- **Interpersonal Intelligence:** The capacity to perceive and distinguish differences in the moods, intentions, motivations, and feelings of others. Includes sensitivity to facial expressions, gestures, and body language. This intelligence also includes the ability to respond to these cues effectively, to work well with others, and to lead.

- **Intrapersonal Intelligence:** The capacity for self-knowledge and understanding, and the ability to act on the basis of that knowledge. Includes having an accurate picture of one's own strengths and limitations, inner moods, intentions, feelings, motivations, needs, and desires, and a capacity for self-discipline and self-esteem.

Gardner's theory states that everyone possesses all seven intelligences, each of us in different degrees and combinations. As he says,

It is of the utmost importance that we recognize and nurture all of the varied human intelligences, and all of the combinations of intelligences. We are all so different largely because we all have different combinations of intelligences. If we recognize this, I think we will have at least a better chance of dealing appropriately with the many problems that we face in the world.

The exact number of "intelligences" is less important than the idea that there *are* many kinds. Gardner himself has built on the core seven. He's discussed, for instance, a "Naturalist Intelligence," saying "...the individual who is able readily to recognize flora and fauna, to make other...distinctions in the natural world, and to use this ability productively...is exercising an important intelligence and one that is not adequately encompassed on the current list." He's also referred to a "Spiritual Intelligence." And he's called the ability to blend or synthesize different intelligences so they work well together a "Synthetic Intelligence."

These ideas have had a significant impact on education. A number of schools around the country are integrating them into their assessment and teaching practices. More and more teachers, when assessing student achievement, are finding ways for students to work within their areas of strength to demonstrate what they've learned. They're also finding ways to encourage students to develop abilities and intelligences that may be more challenging for them. Gardner's work supports findings that **education works most effectively when it takes into account the unique blend of ways in which each child learns, thinks, and feels.**

We're guessing that, as you read the descriptions of each of these intelligences, you see your child in some of them. Does she have a knack for building model airplanes? Does he sing like a meadowlark? Is your child a natural dancer, or peacemaker, or journal writer? Remembering and working with your child's many kinds of intelligence gives you more insight into her capabilities, allows her to express a range of strengths, and recognizes the balance among her achievements. We're not all the same, thank goodness. Why should we test and assess only a few conventional skills?

Children benefit from reflecting on their own learning processes and strengths; research supports this. Having a conversation with your children about their strengths can help them choose appropriate ways to solve problems for themselves. It can also provide you with insights that'll allow you to help them develop to their fullest potential. It's important for your child to learn that everybody's "smart" in different ways, and that recognizing our particular strengths can help us learn more effectively. Also, knowing areas where we're *less* developed can encourage us to become more flexible and learn in new ways. Here are some suggested questions to begin a discussion with your child about multiple intelligences. Explain that these questions are meant to get her thinking about the different ways in which she's smart and about how she learns most easily:

TALKING TO YOUR CHILD ABOUT MULTIPLE INTELLIGENCES

WORD SMART
Do you like to read books and listen to stories?
Do you like word games like Scrabble and Password?
Do you like to tell jokes and appreciate rhymes, puns, and tongue twisters?

NUMBER SMART
Do you like math?
Do you wonder about how things work?
Do you enjoy strategy games, logic puzzles, and brain teasers?

PICTURE SMART
Do you enjoy doing art projects?
Are you good at doing mazes and jigsaw puzzles?
Can you imagine pictures in your mind?
Do you like to draw or doodle?

BODY SMART
Do you like sports?
Do you like to learn by watching or by doing?
Do you like working with your hands (building models, sewing, etc.)?

MUSIC SMART
Do you remember songs easily?
Do you like to play musical instruments, tap out rhythms, or sing?
Do you like to listen to music?

PEOPLE SMART
Do you prefer to invite friends over, rather than play by yourself?
Do you have several different close friends?
Do you prefer talking about problems with someone, rather than working them out yourself?

SELF SMART
Do you like spending time alone?
Do you like thinking about what you want to do in the future?
Are you happy to be a little different from the group, or do you like to fit in?

Remember IQ tests? IQ stands for "Intelligence Quotient," and this test was developed by French and German psychologists (the name was coined by an American) in the early 1900s to measure what was considered a person's mental capacity. The test's many critics claim that it measures only a narrow band of intellectual strengths—mostly "school smarts"—while ignoring other forms of intelligence. Partly because of those limits, it's also thought that these tests are biased against certain ethnic and socioeconomic groups and in favor of others. The IQ test couldn't be more contrary to the idea of multiple intelligences.

Well, here we are, at one of the BIGGEST dilemmas for every parent...

Homework: Pros and Cons

The history of homework in the United States

Educators Etta Kralovec and John Buell, in their book *The End of Homework* (Beacon Press, 2000), describe two revealing poles in homework history. Here are the basics.

In 1901, under pressure to repair what many felt was an emotional and unjustified burden on children, the California Civil Code banned homework outright. By 1930, at the height of national schoolroom and school strategy reform, condemnation of the practice had organized into the Society for the Abolition of Homework. The effectiveness of homework was in grave doubt during this period, and its possible toll on children's health was a pervasive topic.

A few decades later came an abrupt political wake-up call. In 1957 the Soviets launched *Sputnik*, the world's first artificial satellite, and nabbed the first national presence in space. It was like the starter's gun on a racetrack. The United States unleashed a drive to improve math and science education and create a generation of graduates who could put the country back on its competitive feet in space exploration and technology. Homework was increased across the school system. A year later, the National Defense Education Act (NDEA) was created to spur math and science education at all levels. It, too, mandated the assignment of more homework.

In the space of fewer than two generations, homework had evolved from a maligned practice to a national priority.

Today, a fresh emphasis on homework makes the newspapers every day. The widespread publication of statistics—who's learning how well, based on what tests—feeds our fears that our children aren't as well prepared as those in other countries. (Take care with those statistics—see "Look behind the headlines" on page 65.) While there are legitimate problems with our academic preparedness in this country, and our national science literacy and proficiency rates must absolutely be addressed, it isn't at all clear that the mounds of homework our youngsters are assigned is the way to make the improvement.

The arguments against homework

One widely quoted University of Michigan study suggests that six- to eight-year-old schoolchildren are assigned up to three times as much homework as children were twenty years ago. This in a time when, studies tell us, children experience (and need time and energy to process) about 1,000 times more new events per year than they did then! As a parent, you know what it means to see your child labor under a pack of books to and from school, or to see him sit down to literally hours of extended schoolwork at home. You may feel family life slipping away as your youngster's time is swallowed up in isolation or frustration. You may see his other needs—for companionship and free exploration and simple decompression after a long day—go unmet, week after week. *San Francisco Chronicle* writer Katherine Seligman, in an August 2001 article on kindergartners, called this hyperassignment of homework "muscling in" increasing amounts of academic time.

And we don't need to tell you that if your child also participates in organized extracurricular activities (soccer, piano, gymnastics, softball), her time and yours are at an even greater premium.

If excessive homework is a routine problem, speak to the school, starting with the teacher(s). Parents' feedback is important and helps set the homework standard at a school.

In the United Kingdom, the English government is pushing for more homework in primary schools (and trying to prolong the traditional school day, as well). The National Association of Head Teachers (NAHT) in the UK vigorously opposes this. It thinks children should be spending more out-of-school time in extracurricular activities like sports and music, and *less* on homework. After a recent two-year survey, researchers from King's College, London concluded that homework has little impact on raising math achievement in primary school. And a University of Durham study of eleven-year-olds found that children who are given more homework do *not* perform better in tests than those who do very little—and they often perform worse.

Our brains are amazing. They really can accommodate limitless amounts of new information. But they do it best, and longest, when we're rested and in a balance between intellectual concentration and creative variety.

In the long history of educational research, there are **no studies that prove**

that extra homework leads to extra success in school or in life. In the thinking of many parents and educators, the current homework load imposed on our children is unjustified and unhealthy. A parent quoted by the BBC in August of 1999 put it this way: "I have calculated that between the ages of eleven and sixteen, children spend almost 6,000 hours at school. If you can't learn all you need to know in that amount of time, then there's something seriously awry with current teaching methods."

Perhaps the most telling argument against homework is that a lot of it (as usually assigned) fails in its stated aim to advance learning—and in fact turns off, frustrates, or even embitters far too many children. Yes, some tantrums may be a sign of a pending learning breakthrough. But when we're talking about a child in second or third grade who's being subjected to two or three hours of homework every night, it could be argued that the impact on that child's future learning winds up being traumatic and destructive.

In later grades, in middle school or high school, when there's a "ton" of homework in four or five subjects every night, the sheer mental and physical burden on both students and their parents can be overwhelming and exhausting. It can also destroy any love of learning. To add insult to injury, it's the rare school in which teachers are able to consult and coordinate with one another to ease this burden in an intelligent, strategic way. This situation would be at least partially improved if teachers worked together so that major assignments in different subjects didn't coincide with one another on the same night!

The arguments for homework

That said, there are also many proponents of homework, and arguments that say homework strengthens the connection of home to school and parent to teacher. Throughout this book we cite the importance of parent involvement with a child's schooling; and certainly, engagement together over homework is one venue for that participation.

There are parents who feel reassured by school assignments they can see with their own eyes, and who judge the quality of a teacher by the quantity of homework he assigns. (And some teachers—maybe many—feel pressured to assign heavier homework loads by parents who equate lots of homework with a "good education" and a "suitably challenging

We might coin a new term—Family Work—to describe motivating and interesting assignments that purposely involve the whole family...or parents and other adult relatives...or siblings or peers, etc. A third-grader could interview his extended family on topics in many subject areas. A middle-schooler studying why the seasons change could use a multiple-choice survey to find out how well her parents and grandparents understand "the real reasons for the seasons." There are many possibilities for "family work"; assignments that build on and strengthen learning and achievement in school—and have many other benefits besides!

teacher." You can witness this pressure at Back-to-School Nights, when one or more parents vocally complain to a particular teacher: "You're not giving enough homework!") In an article for *USA Weekend* (August 24-25, 2001), educator Tom Loveless argued vehemently that homework is *under*-prescribed in this country; and he's not alone. Loveless takes on proponents of less homework (including the authors of *The End of Homework*), chiding them for inaccurate statistics and challenging the foundations of their claims. What are the most common arguments in favor of homework? That homework instills a sense of discipline in students from an early age, and prepares them for huge amounts of homework in later grades (if that's a sensible objective!); that homework enables students to spend more time on a particular subject than they can in the busy school day; and that homework is a way for students to practice things they need to practice (multiplication tables, math word-problem solving, spelling). After all, many homework supporters would argue, repetition and reinforcement, while not exciting, are necessary for learning.

Clearly, "homework" means different things to different people.

An important homework distinction

While it may seem obvious, it's worth emphasizing that two aspects figure into any discussion about homework: the **amount** of homework and the **kind** of homework. Perhaps if we think again about the different kinds of homework we can shed some light on the debate. There's homework that involves practice of skills—for which many feel there's a place and a need. There's homework that features intriguing problem-solving challenges. And there's the distinction between "make-work"/heavily repetitive homework and a variety of different assignments. Proponents of long-term assignments (things like a researched biography in history, or a science project) contend that this kind of homework fosters organization and discipline, especially when the teacher has stipulated clear step-by-step benchmarks for progress.

The only thing for sure

You know your child better than anyone. You know your home situation best. You know how much work your child brings home. Whether or not you believe in homework, it's a current fact of life. Unless your child is schooled at home, she's probably bringing work home every night. And if she attends a typical North American middle school or high school, it may be a groaning amount.

So here's the issue. Whether or not you believe in homework, how can you best cope with it?

Making the Best of Homework

How involved should you be?

Schools can't know or control the situation a child goes home to every night. The circumstances in some homes make homework much more difficult than in others. That's reality. But as you've now heard us say a dozen times, research has proven that your simplest act of involvement—at whatever level—makes up for obstacles. That goes for homework too. As a start, see how many of the following ideas you can put into practice, right now or over time, to help your child with homework. (For more ideas, we recommend *Helping Your Child Succeed in School*, by Michael H. Popkin et al. [Active Parenting Publishers, 1995], from which we adapted some of these with permission.)

HOW TO SUPPORT YOUR CHILD'S HOMEWORK

SHOW INTEREST. As we've seen, the messages you convey about schooling register deep inside your child and can influence her whole attitude toward education. Homework, as a subset of schooling, is the same; your child knows immediately if you're genuinely interested in what she brings home. So the message here is, take a real interest. Ask questions, care, and be curious.

BE A CONSULTANT, NOT A TUTOR. It's really the parallel of teaching a person to fish rather than putting a trout on her plate. For all that your active interest in your child's homework and the value you place on it are immensely important, there's a potential trap. If, in your enthusiasm, you end up doing much of the work FOR your child, she won't take much away from the experience. Even worse, she'll be learning to surrender responsibility to others. Lack of hands-on participation is a fast track to giving up.

MAKE HOMEWORK THE RESPONSIBILITY OF YOUR CHILD. Your child needs to assume responsibility for her homework. This sounds obvious in theory, but in practice it's hard not to assume the role of "project manager" for certain things—like planning every minute between 3 P.M. and bedtime so there's time for homework and everything else, or breaking down a large project into individual steps and then planning a schedule to accomplish those steps. These aren't skills we're born knowing; they need to be learned. Homework provides a good context for learning these important lifelong skills.

Be aware that certain years mark important transitions in the kinds and amounts of homework assigned to students. As your child hits 6th or 9th grade or other such junctures, make yourself available to help him adjust to bigger workloads or more substantial projects. The greatest contribution you can make is to **help your child learn how to manage these challenges himself.**

Don't hesitate to call your child's teacher to discuss her homework philosophy or ask her advice about supporting your child. Be on hand to help your child set up his systems for success and to adjust to these changes, but make it his job to stick to them.

BE WILLING TO OCCASIONALLY LET YOUR CHILD "FAIL." By living with the consequences of "falling down on the job" on an assignment or two, your child will learn that being motivated and responsible is what it takes to be on time and to get hard things done.

CREATE THE ENVIRONMENT. If you possibly can, provide the right kind of time and space for homework. A designated homework time (always after dinner, for example, or always after school, always before television) provides a reassuring routine, and sets your child up for times in the future when he'll need this kind of need-to-count-on-it self-discipline. Also, as you've probably experienced yourself, we often get subtle "cues'" from familiar environments; a certain light from a certain direction on a certain table can put us "in the zone" for work, making it easier to get started and concentrate.

While some children can more or less block out noise in order to work, a quiet environment obviously lets your child concentrate better. It also adds to her sense that you value her homework time. If you're able, carving out a specific time and place and providing a quiet environment in which she can do her best helps your child see how much you care about her academic efforts—and that you'll do everything you can to support her in them. Again, even a less than ideal place or time, if it's created especially for your child, can send a strong message of support.

BE POSITIVE ABOUT HOMEWORK. It's hard to convey enthusiasm for your child's homework if you think it's too much, too little, too hard, or too frequent, but remember: it is what it is. As the person who expects the best from your child, and who knows both his potential and his limitations, your proactive response to homework can motivate him to perceive it with curiosity rather than dread. Focus on helping your child do things effectively and efficiently as the way to "beat the homework drudge." It's amazing how much time we save by skipping procrastination and complaint and moving directly to a specific task! Not all children can heed this advice (at least not immediately), but demonstrating that you understand that homework isn't always fun but that it needs to get done will make a fundamental difference to your child.

TURN THE SITUATION INTO A LESSON ABOUT SUCCESS IN LIFE. Talk to your child about the value of figuring out what's needed in a situation. Ask her what makes a particular teacher satisfied. Is it different from teacher to teacher? Knowing how to judge a situation and meet its requirements is an important life skill. It's okay to talk about doubts you may have about homework in the context of the lifelong success principle: each situation has requirements that need to be met.

And one more thing:

BE ON THE LOOKOUT FOR BURNOUT. It'll be clear, if you're attuned to your child, if she's feeling maxed-out by her schedule or workload. Think before signing your child up for yet another organized sport or dance lesson or paper route. **You can** cut back on extracurricular activities like football practice or piano lessons. **You can** discourage your child from taking that part-time job. You can hire a tutor, to bring fresh inspiration and support to the situation. And **you can** talk to your child's teacher or the principal about your concerns.

In high school, as your child increasingly feels the pressure of meeting college requirements, you may also want to discuss directly with her how she can best schedule the course load, and encourage frequent consultation with a counselor (if available) to plan this workload as well. Fixating solely on the most advanced courses, the highest possible grades, and astronomical SAT scores can lead you and your child to forget that, increasingly, even the most academically demanding universities are looking for well-rounded students and a demonstrated range of activities, community service, career goals, and social purpose.

Helping your child recognize high standards

Every child wants to succeed, but not all children know what to do to be successful. Learning how to be successful—really, what a person needs to do to get along in life—is a big part of what the growth from childhood to adulthood is about. As a parent, you can help your child acquire her own strong sense of how to be successful. The first step is to recognize what your child is doing that helps her succeed in her work, and what might be holding her back.

QUESTIONS TO ASK YOURSELF WHEN CHECKING YOUR CHILD'S WORK

- Has she answered the question that was asked?

- Does she need to provide a more complete response?

- Do the ideas make sense? Are they presented in logical order?

- Did she provide an explanation of how she solved the problem?

- Did she explain why she knows something to be true?

- Is there information she doesn't have, making the task harder?

- Is she holding onto some inaccurate information that's interfering with her understanding?

A child who understands the standards for good work is more likely to produce good work. By asking himself certain key questions as he works, a child can recognize and achieve high standards even without a teacher or parent on hand. You can help your child learn what to ask himself, and prepare him for his increasing role as an independent learner, by helping him internalize the following kinds of questions:

QUESTIONS YOUR CHILD CAN ASK HIMSELF WHEN CHECKING HIS OWN WORK

- How did I figure that problem out? Can I explain it in words?

- How do I know that's correct?

- Why do I think this? I'll write my thinking down.

- Can both these things be true?

- If I'm coming up with more than one answer (and think there should be only one), does one response look more correct than the others?

- Can I find a better way to persuade the reader of my answer?

- Can I make a drawing that shows what I mean?

- Are my drawings labeled?

- Did I describe the units of measure correctly (using inches, millimeters, teaspoons, etc.)?

- How is this similar to what I did in class?

- What part is hard for me? How could I make that part easier?

These are questions you may also want to use yourself when assessing your child's work. The way your child responds to them provides insight into how she's thinking and learning, and they encourage an internal quest for excellence.

Turning discouragement into encouragement

So your child brings home a science assignment—say, to build a model rocket. She's having a hard time with it, and comes to you for help. "Sorry honey," you say, "I don't think well in three dimensions—and anyway, I always got confused with instructions. Maybe your mother remembers how to build things better than I do."

"A science model?" her mother says, "I've forgotten all I ever knew about building projects—and that wasn't much. And anyway, I always hated astronomy."

Did you consciously decide to discourage your child? Not for a moment. But look at the messages she got:

1. My parents haven't remembered much about science—it can't be that important.

2. My parents never really understood (or even liked) science; why should I?

3. Building models is hard.

4. When I have trouble understanding something, I may as well give up.

A parent can unwittingly discourage her child in several ways. The authors of *Helping Your Child Succeed in School* (Active Parenting Publishers, 1995) have identified the four most common. Here they are, with the authors' suggestions and ours for turning them around:

MOVING AWAY FROM DISCOURAGING MESSAGES

DISCOURAGING: FOCUSING ON MISTAKES.
A whole lot of time and effort can go into a child's paper or project or math quiz. A typical parent "knows" this...but so often it's the three misspelled words or the errors in subtraction that catch your eye—and in that one knee-jerk moment, it's possible to blurt out the mistake without appreciating the child's greater effort.

INSTEAD: FOCUS AND BUILD ON STRENGTHS.
Children, like us, are encouraged and made brave by their successes. Motivate your child. Help him experience the joy of achievement. Point out the strengths he called on to make that achievement happen. Comment on the effort, not just the results.

DISCOURAGING: PERSONALITY ATTACKS AND PERFECTIONISM.
There's no possible advantage to calling a child names. "Stupid," "lazy," or "careless" leaves no room for improvement, and such labels can seriously damage a child's courage and core sense of self. Besides, why should a child work to be "better" than what her parent seems to believe about her? It's a no-win tactic.

INSTEAD: FOCUS ON THE PROBLEM BEHAVIOR, NOT THE PERSON.
Rather than labeling a child's personality, address the activity or behavior that troubles you. Instead of "You're lazy," point out "You didn't do your homework." Instead of "You're careless," try "You aren't proofreading your work."

DISCOURAGING: NEGATIVE EXPECTATIONS.
Children can sense when we expect little from them, even if we don't say it in words. You may *say* "I know you can do this" to a child you secretly think is hopeless in math, but your tone of voice or your body language or some other subtle form of expression will send an entirely different message. Similarly, if you barely give your child a moment to consider a question before rushing in with the answer yourself, it's as good as saying "I don't think you can get this." Even if he can't consciously identify what feels wrong about this behavior, a child can become discouraged and simply stop trying.

INSTEAD: SHOW CONFIDENCE.
Show that you believe your child is capable of pushing ahead even when she's frustrated or feels defeated. Show confidence that she'll eventually succeed. Allow her plenty of time to consider questions. Send the message that you're confident she'll go on to great things with her life. This encouragement can make the difference between success and failure.

DISCOURAGING: OVERPROTECTION.
Stepping in too soon, too often, or too automatically can also send your child the message that you don't think he can handle a situation. An overprotected child will give up easily and demand help in everything he tries. He may suffer fear and isolation when asked to be self-sufficient.

INSTEAD: LET IT HAPPEN.
Allow your child to experience and overcome his own frustration, solve his own problems, and accept the consequences of his own choices (within the bounds of safety, of course). What he learns from these experiences will equip him to succeed in school and in the community. He'll be much more self-sufficient and productive.

The real "work" of encouragement is mostly this: be open to opportunity, and be vigilant about inadvertently discouraging or shutting down your child. By creating a general freedom to try or ask or investigate, you create an environment in which challenges are treated as opportunities, not as prompts for old and anxious ghosts of your own experience.

In Praise of Praise—or PTC-TWB
by LB

I've got a bumper sticker on my car—"Praise the Children and They Will Blossom." It's said to be an old Irish saying, but I expect proverbs like it abound in many cultures. And while there are many lessons every parent learns over time, my experience has taught me there's a whole lot of truth in that message on my bumper sticker.

"Praise the Children and They Will Blossom" (PTC-TWB, for short). Easier said than done, though. One of the most wonderful educators and parents I've ever known told me that sometimes when she sees my bumper sticker she feels a little sad or guilty. Why? Because she's not always able to keep its positive message in mind amidst the hurly-burly of stressful schedules, family dynamics, and the whirling-dervish activity of her active children. I reassured her instantly—she's a great mom! She really does put the proverb into practice. And of course, no one of us is perfect (far from it!); no one's able to practice every aspect of good parenting every day!

Besides, praise is a wondrous balm when it's merited, accurate, honest, and heartfelt, but, like anything else, if it becomes routine, automatic, formulaic, or false, it loses its magical power.

Still, of one thing you can be sure: When a parent sends a child a message of sincere praise, there's nothing more beautiful than the glow that emanates from that child's eyes—as the growing plant begins to bloom. If a plant needs soil, water, and sunlight, then perhaps for a child the "soil" is the home, school, and community environment; the "water" is the needed nutrition/stimulation of all kinds (physical, mental, and spiritual); and the "sunlight" is praise. PTC-TWB.

Many of the smallest examples of our own parents' praise (or criticism) seem to stay with us all our lives—there must be a reason for that! I was fortunate in having parents whose encouragement was authentic and meaningful. I've encountered ultra-critical parents who, even when they praise, can never seem to do so without some reservation. And I have met far too many people whose parents—whether they were attentive and well-meaning or neglectful and severely judgmental—left memories that seemed to belittle, ridicule, or in some other way contribute to an erosion of their child's self-esteem or to cause their child to underestimate his or her abilities in art, or math, or sports, or science. Often this comes in the form of a casual parental remark that's more unthinking than deliberately disparaging. Some parents are more sensitive to their children's emotional landscapes than others—but under stress all of us can get distracted or irritated. It's wise to remember how easily feelings can get hurt and how long those seemingly small negative moments are held inside. When the moments are positive, then, as the song goes, "all the world's aglow"!

One of the moments I remember best is the time I wrote a very early poem about Rip Van Winkle. My mother told me how much she liked it. She had a smile and merry eyes that lit up with eagerness for creative projects, poetry, and art of all kinds—a look that in itself was encouragement beyond words. She had me show the poem to my father, who wrote quite a bit of poetry himself. Part of the poem went like this:

Now Rip, he met some little men,
Not too many, about nine or ten,
They gave him some whiskey and beers,
And then he slept for twenty years.
He grew a beard so very far
It was almost as long as our car...

My father read it carefully, then very enthusiastically congratulated me on what he called my "excellent use of imagery" in the line about his beard being "almost as long as our car." He repeated this several times in the next few days. I was only five or six years old—had no idea what metaphor was—but there can be no doubt that my parents' reactions to my fledgling efforts helped plant a strong sccd of poetry in me!

Some thoughts on computers and the World Wide Web

As the century rushes upon us, new technologies abound. Personal computers more powerful and less expensive than could have been imagined thirty years ago. Multi-task cell phones, vast floods of e-mail, and the ubiquitous Internet are changing our lives.

Yet however "wired" (or wireless!) things may feel these days, the integration of technology into education is far from decided or complete. While many schools are connected to the Web, with computer labs and/or some computers in the classroom, and while there are excellent software and CD-ROM programs, there are still many questions about the most effective ways to use such tools in the school setting. (And lots of research going on to help clarify the educational value and most appropriate uses of interactive computer-based learning.)

It's important to be sure your children have equal access to technology in the classroom, when it exists. The Web publication "Gender Relations in Educational Applications of Technology (GREAT)," out of Stanford University, reports that girls, for instance, generally need much more access to computers than they're currently given. (For more on this, see www.stanford.edu/~ttorres/GREAT/.)

A Department of Commerce survey of household access across the United States (*Falling Through the Net*, October 2000), shows that only half of U.S. households have computers; even fewer have access to e-mail or the Internet. The "digital divide," as it's been dubbed, is very real. This disparity has been shown to negatively affect families that, through poverty, discrimination, or lack of resource, haven't been able to "link up" to the technology

---

Not surprisingly, minorities and families living in rural areas are less likely to be wired than their White or urban counterparts. The Department of Commerce survey found that fewer than a quarter of Black and Hispanic households had Internet access in August 2000, compared to half of White households and almost sixty percent of Asian American and Pacific Islander homes.

---

burgeoning around us. For even quite young children, when other kids in a school seem to be wired at home, it can feel awkward to be left out.

At home, access to a computer and to the Web can have many benefits. If you don't happen to have access, there are helpful alternatives to going entirely without; increasingly so. Some opportunities already exist in the community, some you can create yourself. Your child's school may have computer lab hours. Many libraries now have Internet-connected computers available for use without charge. Community centers, such as the Y, often give computer courses and have "public hours" at little or no cost. You might try to form a parent's club or computer pool with other "non-wired" parents in your neighborhood, to work with neighbors who do have access. Try your church, club, neighborhood adult school, and any other organization you can think of. With imagination and a few visits or phone calls, you may find access pretty close to home.

THE BENEFITS OF TECHNOLOGY TO YOUR CHILD'S SUCCESS

- **It lets your child lead the way.** As "grown-ups" well know and often joke about, many children and teenagers have extraordinary computer skills—this can be a case where they teach you! Exploring a topic together on the Internet, or going together to obtain an educational software program and taking time to work with it together, can be a great interaction, and has benefits far beyond the actual content of the program.

- **It broadens your child's access to information.** In the course of this book we emphasize a special way to think about science education and your child's success: inquiry-based learning. The Internet is a fantastic tool in the service of inquiry. One of the things real scientists and student scientists (as well as social scientists) do is research the work of others who've pursued the same or related questions.

- **It teaches your child new skills.** Access to the Internet has revolutionized our access to information. Learning how to do research on the Internet, and especially how to sift the wheat from the chaff, is a very helpful lifelong ability. Educators have noted that the nonlinear process of searching a web of related (but not necessarily physically connected) information takes patience, and must be learned. That's skill acquisition—and it's a different skill from those needed for other kinds of tasks.

All this aside, **while there are important benefits to computers and Web access, they are not essential to a child's academic success.** There are ALWAYS alternatives to searching the Net or typing up reports on a computer. Technology is less important than being mentally engaged—and that, you and your child are supremely equipped for!

Do Boys and Girls Learn Differently?

Males and females are equal in their common membership of the same species, humankind, but to maintain that they are the same in aptitude, skill or behavior is to build a society based on a biological and scientific lie.
—Anne Moir and David Jessel; Brain Sex: The Real Difference Between Men & Women *(Dell Books; reissue ed. 1993)*

The big picture on gender differences

While every child is unique, some generalizations can be made about boys and girls on the whole.

Differences between males and females include not only those that are physically visible and obvious but also a wide range of developmental, structural, and functional differences in the brain. One example is that girls' brains mature earlier than boys' brains, a difference that starts at birth and continues until the brain has myelinated (become "hard-wired," or "set") in our early twenties. And boys' brains have less-active left hemispheres, where verbal and abstract thinking occurs most.

These differences can account, in part, for girls' tendencies to have better verbal abilities (including reading, writing, and a heavy reliance on oral communication), and boys' tendencies toward nonverbal communication and better spatial abilities (such as those used in measuring, mechanical design, geography, and map reading). These are just two of many generalizations made about the relative tendencies of males and females, ranging from differences in fine and gross motor skills to the ability or desire to express feelings and a sensitivity to sensory input.

This *doesn't mean* that all girls are more articulate than all boys, or that all boys are better spatial reasoners than all girls! It means that each group *tends* to perform better in different kinds of tasks and contexts. These differences aren't a judgment of the relative intellectual power of boys and girls, but a generalization about their intellectual tendencies.

One of the dangers of generalization is that it can become a self-fulfilling prophecy. It's critically important to remember that individual children defy these generalizations all the time! Many brilliant writers are men; many brilliant architects and scientists are women. Your child is not an "average," and may not follow these gender norms.

The chicken or the egghead?

The fact that there are brain differences doesn't address *why* these differences exist. Some researchers are quick to conclude that because these differences are physical, they must be innate, or "built in." And to some extent that's probably true. However, there's much evidence to show that behavior *influences* brain development, so others argue that just engaging in certain kinds of activities creates or accentuates these brain differences—and that therefore children's skills and behaviors are anything but preset. That's probably true too. Think about a parallel example related to body development: It's undeniable that boys (on average) are built with greater muscle mass than girls. But all you have to do is look at the increased number of fit, strong girl athletes in the current soccer generation to see how opportunity creates outcome. If you were to measure the gross motor abilities of these girls over time, you'd be able to demonstrate that they developed greater skill and ability through their involvement in sports. This is similar to what can happen in the brain.

While the relative power of nature over nurture (as this age-old debate has been called) is unlikely to be resolved soon, the cause of gender differences is probably less important to parents than how to deal with these differences. The fact is that children are different from each other, with their own individual strengths and shortcomings. In order to approach every child as a unique learner, it's necessary to acknowledge that there are learning and aptitude differences in humans that are sometimes, but not always, associated with gender. (See "Your child's learning style" on page 71.) The question is how to ensure that every child has the opportunities to succeed in any of the wide range of academic arenas that make up our world. It's knowing how to provide for a child according to his or her individual needs that's important.

How gender trends affect your child's academic success

As a parent, it's important to know the possible pitfalls for your boy or girl (again, speaking in general terms) so you're better prepared to provide your child with every possible advantage. In *Boys and Girls Learn Differently!: A Guide for Teachers and Parents* (Jossey-Bass, 2001), Michael Gurian and Patricia Henley broke new ground. What had been intuitively believed by many educators for years, that boys and girls present different learning challenges, Gurian's team confirmed in a two-year focus study.

Several points in these tables are adapted with permission from *Boys and Girls Learn Differently!: A Guide for Teachers and Parents,* by Michael Gurian and Patricia Henley with Terry Trueman. © 2001 by Michael Gurian. Used by permission of Jossey-Bass, a subsidiary of John Wiley & Sons, Inc.

ADVANTAGES FOR GIRLS, POTENTIAL DISADVANTAGES FOR BOYS

- **Extracurricular activities.** Girls make up the majority of student government officials, after-school club leaders, and school community liaisons.

- **Academic performance.** Girls choose to take harder courses in middle school and high school than boys do; they also study harder than boys, on average, for all courses. Girls receive approximately sixty percent of the As, and boys receive about seventy percent of the Ds and Fs. Of students performing in the top fifth of high school grade ranges, sixty-three percent are girls.

- **Specific academic areas.** Girls are approximately one and a half years ahead of boys in reading and writing competency, according to statistics tracked by the U.S. Department of Education since 1981. This advantage exists at all levels of ability, not just the highest.

- **Educational aspirations.** Our college populations are now sixty-percent female. The Department of Education has found that eighth-grade and twelfth-grade girls have, on average, higher educational aspirations than boys. A nongovernmental study corroborated this finding: three-fourths of girls (compared to two-thirds of boys) "believe that they will have many opportunities available to them after they graduate" from high school.

- **Learning and behavioral health.** Females are less likely to experience certain learning, psychiatric, or behavioral disorders than males. Boys make up two-thirds of learning-disabled and ninety percent of behaviorally disabled young people. They account for nearly 100 percent of the most seriously affected. Boys constitute eighty percent of ADHD (attention deficit and hyperactivity disorder) and ADD (attention deficit disorder) diagnoses, and suffer seventy percent of serious drug and alcohol problems. For every girl who actually *commits* suicide, four boys do.

- **Discipline problems.** Girls present fewer discipline problems and are less likely to drop out than boys. Boys account for ninety percent of the discipline problems in school, as well as eighty percent of dropouts. Boys garner the majority of school punishment for immature behavior and leave school at a higher rate than girls.

ADVANTAGES FOR BOYS, POTENTIAL DISADVANTAGES FOR GIRLS

- **Athletics.** The majority of sports funding and community support still goes to male athletics in most U.S. schools. Only thirty-seven percent of high school athletes are girls.

- **Performance on tests.** Boys score slightly higher than girls on SAT and other college entrance exams. (This is true even in the verbal categories; boys seem to focus on the core questions more quickly than girls, and respond better to the multiple-choice format of SATs.)

- **Learning and behavioral health.** Boys make up fewer than half as many cases of clinical depression as girls. For every boy who *threatens*

> to commit suicide, four girls attempt it. Boys are less than half as
> prone as girls to eating disorders.
>
> • **Specific academic areas.** Boys are two to four points ahead of girls in
>   math and science scores tracked by the Department of Education,
>   particularly at the highest levels of calculus, chemistry, and physics.
>
> • **Classroom behavior.** Boys tend to be louder, more physically
>   aggressive, and more prone to attention-getting behavior in class-
>   rooms than girls. As a result, boys take up more teacher time.

Gurian and Henley stress that cultural expectations can play a considerable role in the diagnosis of learning "disabilities." In our culture, for instance, where early reading ability is considered a hallmark of achievement, boys (traditionally less verbal than girls) may get tagged "learning disabled" when in fact their brains just need more time to catch up in reading and writing.

To the last list, "Advantages for Boys," we might add teacher bias. In a lin-gering historical imbalance, teachers tend to direct more questions, allocate more computer time, and otherwise pay more direct attention to boys than to girls in a classroom (even without overt attention-seeking from the boys). Girls may have had access to formal education in this country for a couple of hundred years, but there's a long history of discrimination. We need to make much more progress—in society and in the classroom.

Surprising finding #1: in school, our culture favors girls

At present, the educational system and individual classrooms are generally designed in ways that work better for girls than for boys. Part of this is a tribute to the efforts and mobilization of educators and feminists in the last couple of decades to address girls' lack of access to math and science and higher education. The consequence is that young women now outnumber men at every level of higher education, and girls' lower achievement in math and science has all but been eliminated; it now exists only in the very highest echelons of math and physical science. These gains are a triumph.

The challenge now is to provide learning environments that work as well for boys as for girls. Boys make up the majority of "challenged learners," receive the most failing grades, and drop out of school in the highest numbers. Our educational system is not serving these students well. For one thing, school requires a lot of sitting and listening—something that's harder for kinesthetic learners (people, mostly males, who learn best by moving around and getting their hands into things). For another, our nation's K–12 schools and classrooms are largely led by women, and research has shown that boys generally respond better to male mentors.

Interviewed by Michelle Galley in *Education Week* (January 23, 2002), Harvard education specialist William S. Pollack said this:

> The curriculum is set up in a way that works better for girls. Not because we are favoring girls, [but] because it just happens to work better for girls, and the boys don't do it as well.... [Then] we say boys aren't performing well enough, and they need special attention. Now that would be okay if special attention really meant paying attention to special needs. But we all know that "special needs" is really just a euphemism for being to some extent defective and not fitting into a system.
>
> Erik Erikson, the great child psychologist, said years ago that when a disease or a disorder in children reaches a number above ten or twenty percent, we need to stop calling it an illness or a problem and start saying it is normal or normative, and wonder what it is about our society that is wrong, rather than the child.

(For more on how the classroom environment affects boys and girls, you may wish to read *Real Boys: Rescuing Our Sons from the Myths of Boyhood*, by William S. Pollack; Owl Books, 1999.)

We must acknowledge the diverse needs of our diverse students and *truly* leave no student behind.

(Less) surprising finding #2: in the real world, our culture favors boys

Girls' greater success in school doesn't necessarily translate to greater advantage out in the world or in the job market. While college graduation is the most consistent indicator of stable future income, and women have achieved more stable income than in the past, our male-dominated culture is slow to relinquish power. Men still make more money (compared to women in general and to women in comparable jobs) and occupy the most powerful and influential jobs in the United States.

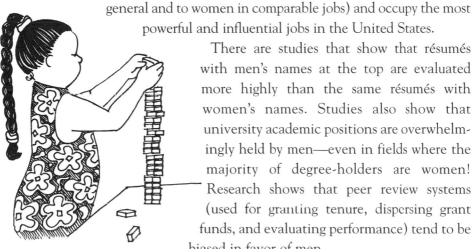

There are studies that show that résumés with men's names at the top are evaluated more highly than the same résumés with women's names. Studies also show that university academic positions are overwhelmingly held by men—even in fields where the majority of degree-holders are women! Research shows that peer review systems (used for granting tenure, dispersing grant funds, and evaluating performance) tend to be biased in favor of men.

Our society is far from gender-bias free.

How Far Have Women Come?

- Roman philosopher Musonius Rufus (c. A.C.E. 30–101), when asked if sons and daughters should be given equal education, pointed out that people who worked with horses or dogs made no distinction between male and female in their training. (Well it was a start, anyway.)
- The first school for girls in this country opened in January, 1792. The Young Ladies' Academy of Philadelphia taught young girls reading, writing, spelling, arithmetic, grammar, geography, music...and catechisms.
- In 1798, Charles Darwin's grandfather, Erasmus Darwin, wrote "Plan for the Conduct of Female Education," a list of resources available to instruct young girls in "Music and Dancing, Reading, Arithmetic, Geography, History, Rudiments of Taste, Arts and Sciences." (Also "Morals, Beauty and Grace, Drawing, Prudence, Justice, Chastity, Religion, Dress, and Stampering," but what can you do.)
- The first college for women in Illinois, the Jacksonville Female Academy, opened in 1830.
- Mary Fairfax Somerville, a British mathematician, published *The Mechanism of the Heavens* in 1834.
- In 1849, Elizabeth Blackwell became the first woman doctor in North America.
- Marie Curie won the Nobel prize in chemistry in 1911, for the isolation of pure radium.
- The first African American woman pilot in the world to be licensed was Bessie Coleman, in 1921.
- Aircraft manufacturer Olive Ann Beech cofounded Beech Aircraft in 1932.
- In 1933, Frances Perkins was appointed Franklin Roosevelt's Secretary of Labor— the first woman to hold a cabinet position in the United States.
- Sri Lanka elected the world's first woman head of state, Sirimavo Bandaranaike, in 1960.
- Colonel-Engineer Valentina Tereshkova became the first woman in space in 1963, aboard the Soviet spacecraft *Vostok 6.*
- Dr. Sally Ride, in 1983, became the first American woman to orbit Earth when she flew aboard Space Shuttle *Challenger.*
- Dr. Sethanne Howard became program director for the National Science Foundation's Extragalactic Astronomy and Cosmology Program in 1988.
- By 1997, Secretary of State Madeleine Albright was the highest-ranking woman in the history of the United States government.
- In 2001, Dr. Julie Payette was appointed Chief Astronaut of the Canadian Space Agency.

How you can help level the playing field

Here are some concrete suggestions for maximizing the advantages for your child—boy or girl:

HOW TO LEVEL THE FIELD FOR YOUR GIRL OR BOY

- Learn your child's strengths and challenges.

- Make sure your child has a broad variety of experiences. Don't let social pressures or your child's individual preference narrow his or her opportunities. For instance:

  Ensure that boys (and girls who don't match the typical gender-type) have opportunities to:
  - Learn and practice fine motor skills (like sewing, beadwork, calligraphy)
  - Combine reading and writing with related active and hands-on projects
  - Articulate and expand on their ideas orally and in writing
  - Express emotions and make personal connections in school
  - Develop, practice, and be rewarded for nurturing behaviors
  - Engage in physical movement as they learn
  - Benefit from mentors of their own gender

  Ensure that girls (and boys who don't match the typical gender-type) have opportunities to:
  - Learn and practice gross motor skills (through athletics and movement games)
  - Use concrete manipulatives (apparatus or tools), especially in science and math
  - Develop competence in technology, computers, and the Internet
  - Match abstract learning activities (especially in math and science) with journal-writing activities
  - Discuss their ideas with other students in small groups
  - Hear about and experience images of competent girls and women in literature, history, and current events
  - Benefit from mentors of their own gender

You Wouldn't Let a Reluctant Reader Not Read....
by JB

It's hard not to fall prey to social and cultural gender expectations. And when we do, it limits the development of our children A friend of mine followed her first daughter's lead: team sports, no way! Ballet's what

she was interested in (not surprisingly, since that's what her friends were doing). By the time her second girl came along, after a boy in between, my friend had learned from her son's experience that team sports and the opportunity to develop gross motor skills, coordination, teamwork and cooperation, game strategy, and all the lessons of winning and losing gracefully were too important to miss. So her second daughter got signed up for soccer—and benefited from the many life lessons it had to offer.

As for me, I bought my first son a doll. Well, many dolls, because he wasn't particularly interested the first time. Whenever I saw a glimmer of interest, I'd try again. Why? Because in the same way that my friend's daughters might not encounter as many opportunities to develop typical "boy" skills, I knew that my son was unlikely to have the same opportunity to develop, practice, and be rewarded for the typical "girl" skills of nurturing, empathy, and emotional expression. I persevered. What's fascinating is that over the period of the first four years of his life, my son flip-flopped between engaging with and ignoring those dolls. Had I stopped offering the opportunities because I'd concluded he wasn't interested, I'd have missed the periods when he was thrilled to have them!

When the peer pressure and teasing inevitably set in (it happens so early!), my son, on his own, shifted from dolls to stuffed animals. Even now, as a twelve-year-old boy, he loves his stuffed animals. His love of stuffies established a positive and gender-neutral culture in our family as his two younger brothers, in the most natural way, have adopted and cared for a wide range of animal babies. I'm sure it's no coincidence that my two older boys are confident and comfortable as they care for their littlest brother.

The lesson? Your child's apparent lack of interest in an area shouldn't deter you from ensuring that he or she has the opportunities to develop a full range of life skills and abilities. Our society understands this when it comes to things like reading; a reluctant reader wouldn't be allowed to choose not to learn how to read! In the same way, we should ensure that girls and boys have the chance to develop important skills that relate to life success, irrespective of gender trends. How many times have you heard a parent say that his daughter is simply not interested in construction toys like Lego or blocks, or that his son isn't interested in art? All the more reason to afford them the opportunities!

The best educational opportunities aren't always born of choice and comfort level. Make sure your child's school doesn't allow "free choice" time to mean that girls never venture into the construction and strategy game centers, or that boys stay away from small-motor and verbal-expression activities. You wouldn't let a reluctant reader....

CHAPTER FOUR

THE SPIRIT OF INQUIRY
Keeping the Scientist Alive in Your Child

The Beauty of Inquiry

The object of all serious research can only be to make two questions grow where only one grew before. This is inquiry at its best.
—Economist Thorstein Veblen

Introducing inquiry-based learning

To inquire is…to ask.

Questions explored through literature and philosophy make us richer human beings, stretching our boundaries and challenging our assumptions. Questions that probe mysteries of the natural world set the stage for all of science! What makes clouds stay up in the sky? How do mirrors work? What are fossils? Why do weeds grow better than strawberries? How is it that butterflies can fly? What can you do with a magnet? Inquiry starts with observation, curiosity, and exploration. It's guided by logic, imagination, and experimentation. And it's illuminated by discovery, understanding, and learning. Inquiry-based learning taps into how your child learns best: investigating and discovering things for herself, as *well* as reading and being instructed. It helps your child develop the habit and confidence for lifelong inquiry, and gives you a new perspective on how she learns, what turns her on, and what she thrives on. (You too, for that matter.)

See page 5 for a summary of what the *National Science Education Standards* have to say about inquiry.

Dr. Glenn Seaborg, the late Nobel Laureate and Chair of the Lawrence Hall of Science, wrote this about inquiry:

> It's a wonderful word, embodying the essence of questioning in science and mathematics. Sowing questions. Pursuing curiosity. Finding out more. Inquiry means looking into something—from the Latin, *quaerere,* to seek, to ask….Inquiry-based science begins with a question, spurs curiosity, invites investigations. Inquiry science empowers students as full-fledged doers and thinkers, opening up the wonders and mysteries of the natural world and known Universe.

In the same way that teachers are now being asked to encourage the development of inquiry abilities in all their students, the role of a parent is to nurture that same spirit of inquiry whenever and wherever possible.

Your little scientist

Who asks the best (and most!) questions? Who notices tiny details others take for granted? Who has the patience to fiddle with something again and again until she figures it out? Chances are it's a four-year-old you know—either that or a scientist. It's not a coincidence that in our earliest years we use a very effective process for learning about the world.

From the day an infant is born, there's so much she has to learn about her world. Getting her eyes to focus; discovering her own hands ("Oh, those belong to me!") and figuring out how to grasp an object; exploring her voice and how to make and mimic various sounds.

As a baby orients himself and begins to gain control over his body, these explorations lead to simple experiments. What'll happen if I drop this rattle? Will a crayon write on carpet? on the wood floor? on my nose?

In time, as babies learn simple cause-and-effect relationships, they begin to make their first simple hypotheses. The stove is hot. There's milk in that bottle. I'll bet that grown-up will chase me if I crawl over to the top of the stairs. And of course—sometimes to our chagrin—they test each hypothesis.

A child must evolve a process for finding out about her world, and interestingly enough, that process is very similar to the process scientists use to find out about the natural world. Observing, comparing, sorting, questioning, experimenting, building and testing hypotheses, and wondering—the parallels are striking! Alison Gopnik, Andrew Meltzoff, and Patricia Kuhl, who've done considerable research on learning and language acquisition in infancy, leave little doubt; they titled their recent book *The Scientist in the Crib.*

The Scientist in the Crib: What Early Learning Tells Us About the Mind, by Alison Gopnik, Andrew N. Meltzoff, Patricia K. Kuhl; HarperPerennial, 2000.

Unlike his sister before him, who'd started asking questions almost from the womb, Angus was the strong and silent type. He didn't utter a word for years. On a balmy Northern California evening two days after his fourth birthday, Angus spoke for the very first time. Not to his parents or me, sitting with him in the garden, but to a bagel crumb on his finger. And not "Dadda," or "All gone," but "Why doesn't the Moon fall into the garden when the lights go out?" —NP

DO BABIES KNOW MATH?

In the minds of many educators and psychologists today, pioneering Swiss philosopher and psychologist Jean Piaget "set the bar" for the study of early-childhood development. Although not an education specialist himself, Piaget's years of intense observation of how, what, and when children begin to learn turned the worlds of teaching and child psychology on their heads. Piaget concluded that the mind of a developing child had a lot more to tell us than we'd ever realized. In the years since his work, new research has both built on and differed from his findings. More recent research indicates that real reasoning begins much earlier even than Piaget and his interpreters have suggested.

No kidding.

"Mathematical thinking begins in infancy," says professor of education Prentice Starkey. "It's as natural a process as language; it may even be innate." With funding from the U.S. Department of Education, Starkey and research psychologist Alice Klein, both of the University of California at Berkeley, have created an early math intervention study called the Berkeley Math Readiness Project. A fascinating component of the project is to pin down the earliest age at which mathematical thinking begins and figure out what can be done to support its development.

Infants are not, as some have suggested, born with a blank slate of a brain. They come into this world with (at very least) a predisposition to grasp certain concepts immediately. Even very young children can assimilate and understand fairly sophisticated mathematical concepts such as ordinal numbers (first, second, third), not just the rudimentary 1, 2, 3 of *Sesame Street*. (In fact, Starkey's team feels that this popular children's show lags behind the research; he expresses frustration that an educational program as widely watched as *Sesame Street* may be missing a valuable opportunity to expose children to more complex math.)

Research shows that when children start acquiring informal math knowledge in infancy, based on concrete objects, it lays the foundation for more formal, abstract mathematics learning in elementary school and beyond. Far from being a gradually acquired, not-for-everyone talent, the ability to "think in math" is literally born in all of us.

Part of being human is to think mathematically.

If your computer supports QuickTime software, you can view a Web video of the Berkeley Math Readiness Project in action, on the site of Science Education Today. Go to www.ucop.edu/science-today/, click "video" in the margin, and find the BMRP item on the page.

What happens to our early inquiry abilities?

In as little time as the first few years after kindergarten, in most school

ENCOURAGING YOUR YOUNG CHILD TO THINK MATHEMATICALLY

1. EXPOSE YOUR CHILD TO MATH FROM SEVERAL ANGLES.
   - **Use concrete and varied materials:** blocks, puppets, crackers, pacifiers, stuffed animals....
   - **Vary the activities:** counting, measuring, combining, pacing off, "taking away" and "adding"....
   - **Make use of several learning approaches (see "Your child's learning style" on page 71):** Find opportunities to bring up mathematical concepts using objects, paper and pencil, and the computer. Count together the number of steps in your child's stride, or the number of ups and downs on a swing.
   - **Be flexible, to accommodate a range of learning:** Grab opportunities whenever they present themselves. Be willing to change course midstream and explore a different topic if your child takes off in a new direction.

2. ACCEPT AND ENCOURAGE YOUR YOUNG CHILD'S "INVENTED STRATEGIES" for thinking mathematically—counting on his fingers, insisting on sharing objects equally (or unequally!), etc. It's been shown that children come up with such devices on their own, as ways of capturing and representing mathematical concepts.

3. DON'T UNDERESTIMATE WHAT YOUR CHILD CAN ABSORB. Ordinal numbers, as an example, are perfectly graspable by even very young children. Find ways of introducing the idea of ordinals (first, second, third, etc.) when playing or in simple conversation with your child.

4. "THINK IN MATH." As you work on enriching your child's mathematical environment, pounce on every opportunity you can think of. Here are some kinds of things you do routinely and some math questions you can introduce with them:
   - **Preparing a meal:** How many cups of water? How many fish sticks?
   - **Setting the table:** How many plates, napkins, glasses? Are there enough (or too many) chairs at the table? How many?
   - **Building a bookshelf:** How many sides does it have? How many shelves?
   - **Parking a car:** How many cars are in front of us? How many in back? Do you think the space is big enough for our car?
   - **In the playground:** How many rungs on the rope ladder? How many kids fit in the fort?
   - **Shopping:** How many people are ahead of us in line? How many pennies have I got?
   - **On the way to a family reunion:** If each person gives a hug to each other person at the family gathering, how many total hugs will there be?

Given the opportunity and provided a math-enriched environment, all children can learn to think mathematically as a matter of course.

settings, children learn that there's often a much greater value placed on "getting the right answer" than on asking a good question. A child's astute observation or episode of profound wonderment may be dismissed as inappropriate by a harried or inexperienced teacher. Soon enough, children begin to give up some of their natural curiosity. They begin to focus on "doing what they're supposed to" in school—which, unfortunately, often involves regurgitating "right" answers and saying what's expected of them. In fact, a typical characteristic of students who do well in school is an ability to give back what the teacher wants.

For teachers, the constraint of working with groups of thirty-some children at once also contributes to squelching some of the individual investigation and expression that can occur in a school situation. Ironically, wondering, questioning, investigating, and exploring are often the casualties in schooling.

Even a child lucky enough to have excellent teachers in the early school years—teachers who conduct open-ended science and math investigations and de-emphasize parroted or memorized "correct" responses—can still close up intellectually. His peers may laugh at "dumb questions," or make fun of what they take to be a lack of knowledge. It takes a will of steel—or a confidence instilled at home—to keep asking questions when your acceptance by the group seems to be at stake.

Keeping the ball in the air

So we know that this natural ability for independent inquiry often begins to unravel when children start their formal schooling. The tragedy is that this is the age at which children are most open to science and math.

As a parent, there's a lot you can do to keep the scientist alive in your child. Several ideas appear below for spotting opportunities for inquiry learning at every curve in the road, and many of the suggestions and anecdotes that abound throughout this book sparkle with inquiry possibilities.

Probably the single most important way to help keep the scientist alive in your child is through your own attitudes. If you approach the world with wonder and curiosity, so will your child. If you ask questions, and make observations and inferences about your environment, your child will pick up the same tendencies. You can model that it's okay not to know something—that in fact it's fun to find out. Demonstrating these behaviors to your child will affect how she approaches school tasks as well as how she views the world. It will help your child develop the confidence and habits of mind for lifelong inquiry.

Create the opportunities at home

Happily, the principles of inquiry-based learning, far from being confined to a classroom, apply in any setting. So while a teacher may need to present a specific curriculum or enrichment lesson in the classroom, inquiry-based learning techniques can be used by any caregiver of a young person or small group of children, anytime.

The ideas that follow are just a seed kit. You'll instinctively and naturally come up with variations that take advantage of your situation, your child's personality, and the natural opportunities unique to your life together.

HOW TO ENCOURAGE INQUIRY IN YOUR CHILD

- **CHOOSE TOYS THAT INVOLVE YOUR CHILD IN OPEN-ENDED EXPLORATION OR INVESTIGATION.** Go for magnifying lenses, shovels, flashlights, blocks, journals, science kits, or modeling clay, rather than video games and other "finite" amusements. Let your child disassemble an old-fashioned alarm clock (and find the best brass spinning tops inside!), or tear open a vintage golf ball for the reams and reams of rubber band and the Superball in the center.

    There are also many home science kits (see "Resources" on page 138) and strategy games (Mancala, tic-tac-toe, nim, chess) that offer marvelous math and science learning.

- **CHOOSE EDUCATIONAL COMPUTER GAMES.** The peer pressure is great, especially for boys, to engage in addictive, mindless, often violent computer games. There are LOTS of educational and fun computer alternatives; seeking them out is well worth the effort.

- **ENCOURAGE YOUR CHILD TO START A COLLECTION.** Rocks, buttons, plastic dinosaurs, pressed flowers, fossils, bones, or other cool things from the natural world—just about anything your household tolerances will allow—can become the basis for your child's sense of mastery in a subject. Becoming "expert" about certain objects promotes fine observation skills, depth of knowledge, confidence, and pride.

- **BROADEN YOUR SELECTION OF FAMILY OUTINGS.** Include science centers, zoos, museums, planetariums, hikes, and nature walks. Observe the night sky together.

- SEIZE LEARNING OPPORTUNITIES IN THE COURSE OF DAILY LIFE. Take an extra five minutes to study the caterpillar on that leaf; count the spots on a lady beetle; study the frost etchings on a window in winter; let your child put his hands in the flour the next time you bake; show her how a bicycle gear works, or an egg beater; observe ant trails when they invade your kitchen; involve your child in cooking or repair projects that require measuring and mathematics.

- FOLLOW YOUR CHILD'S CURIOSITY BY PURSUING THE QUESTIONS SHE ASKS. Have a conversation the next time your child peers at the Moon through a car window and says, "It's following us!" or asks questions about Earth's shape, or about gravity. You don't need to know the answers—in fact, seeking them out by looking in books, searching the Internet, or asking someone who knows is a powerful way to model "how you find out."

    Set out to investigate a question your child has. What will deter an ant trail? What floats in the tub? What kind of bread molds first? What foods do your backyard squirrels like best? What percentage of cars that drive by are the color red?

- ENGAGE IN FUN LEARNING ACTIVITIES IN THE COMMUNITY. Schools often offer Math or Science Nights for families. Most libraries have books filled with safe, easy, and unintimidating experiments that any parent can set up at home. Community organizations such as science centers, the Y, and church groups often offer after-school and weekend hands-on science, math, and computer classes. Especially if science isn't your thing, consider enrolling your child in these classes. See "Resources" on page 138 for ideas, books, and Web sites that can help.

Schooling may occur on schedule, but learning rarely does. As a parent, you have the chance to be spontaneous and follow your child's interests. You and your child will thrive on spending this time together. And if you're not comfortable in your own inquiry abilities, take advantage of some of the many inquiry-based activity resources for parents these days; they're very good and very plentiful.

One of our educators at the Lawrence Hall of Science shares this anecdote from her childhood. Its great charm, and its lesson, lie in the sense we get of genuine discovery in otherwise ordinary moments:

My father had a workbench in the garage. He did carpentry and metalwork, and fixed the family car himself. He often welded and melted various metals, dipping hot metal in cool water and causing noisy steam to come billowing up, volcano-like. My brother and I always seemed to be at his side, handing him things as he asked for them.

I learned a lot by just being there; watching, helping. Maybe that's where a sense of confidence in hands-on, practical things begins. My father showed me how to do things and, I guess, sometimes let me try stuff myself. He was good at explaining what he was doing, and at asking questions about what was going on (Why is the fire blue?). I learned to sharpen a knife (Why does it get hot when you sharpen it? Why use the oil?). I gained a practical sense of tools and how they extend the human hand. His questions, like the questions of a good teacher, spurred my curiosity. They opened my mind across all the scientific disciplines, and they were fun and natural—all within the workaday world of a garage workbench. Maybe that's why I'm so comfortable with science today.

This kind of childhood experience represents the possibilities in any family. It can happen in the kitchen, at the supermarket, in the garden, at the park—anywhere a parent encourages her child to explore, ask questions, and think about math and science in real-world ways. What jumps out in this story is the general culture of curiosity one father created outside of school, in everyday life.

How to Nourish an Inquiring Mind

Foster an environment that values curiosity and investigation

In Chapter Three, under "Turning discouragement into encouragement," we offered examples of unintentional negative messages about school and how they can undermine a child's attitudes about the entire academic process. It bears repeating here: How you express your feelings about science and math has a huge impact on the conclusions your child draws herself. How could it not? Children absorb your attitudes. Young children make your opinions their own. Your subtle and not-so-subtle messages (about issues like schooling or race relations or just what vegetables you don't like) collect in your child in varied and often profound ways—either because he wants to be like you or just by sheer osmosis. There's awesome power in your attitude. It's a little scary, but also wonderful. You can bathe your child's world with positive examples. You can rock that world with contagious excitement for discovery. You can guide her learning with enthusiasm.

WAYS TO CREATE AN INQUIRY-FRIENDLY ENVIRONMENT

- APPROACH THE WORLD WITH CURIOSITY and your child will too. If you ask questions and make observations about your environment (especially if you don't know the answer!) your child will learn that this is a natural thing to do. He'll learn that it's not only okay, but great, to INQUIRE!

  Try setting a dinner-time tradition of asking each family member to share a question she asked herself that day, or something he wondered about. Over time, children will start to pay attention to and value their own questions. They'll learn that curiosity is valued.

- SET THE EXAMPLE. By your everyday actions, as well as by actively creating directed opportunities for your child, you can provide the canvas for investigation, exploration, and discussion. Encourage independent thinking. Share your own questions. Be open to all inquiries. And—this is important—be willing to make mistakes, so your child will too. Self-censorship for fear of being "wrong" stifles learning and discovery in the most damaging way. (Many of us learned that the hard way; "right" or "wrong" is all we were taught.) It's not just about letting go of perceived mistakes (although that's healthy too), it's about reorganizing how we think of inquiry—not as a means to one right answer, but as a gateway to a rainbow of possible responses...and to more inquiry.

- CONVEY A MESSAGE OF DISCOVERY. Wherever in the spectrum of enthusiasm about math and science your child now falls, you can introduce, sustain, or reignite her motivation and confidence. Even the most listless math or science student can be inoculated with a joy for inquiry. Be generous in your responses. Beware of discouraging behavior such as the examples on page 84.

- ASK FOR YOUR CHILD'S PERSPECTIVES. There's magic in being asked what we think, so let your child teach you!

- CHOOSE YOUR REWARDS. The longest-lasting rewards for learning—and for learning to learn—are intrinsic; they come from the inside. When your child educates herself for the sheer love of it, she carries that internal springboard everywhere and forever. Rewarding your child with cash or treats for doing his homework or getting certain grades, on the other hand, may create short-term incentive...but what happens when the payoffs end or the ante goes up? And what message do these extrinsic payoffs send about the value or rewards of education for its own sake? It's a little like giving that person a fish again, instead of teaching her to fish. Instilling a love of learning equips your child forever; pursuit of knowledge becomes its own, self-perpetuating reward. Nurture in your child a genuine hunger for discovery. The academic effects will be broad, long-lasting, and inventive.

In reference to the 10,000 times he failed in his storage battery experiments, Thomas Edison said "Why, I have not failed. I've just found 10,000 ways that won't work."

Reframe the questions you ask

The way a question is asked can either shut down a conversation ("Did you notice the web on the window this morning?") or encourage an exchange ("What did you notice about the web on the window?").

Open-ended questions ("Why do you think the butterfly flew away?") encourage creative thinking better than closed questions ("How many antennae does a butterfly have?"). Both kinds of questions, those that elicit straightforward information and those that provoke more complex thinking, have a place.

When your child asks a question, one good first strategy is to turn it around and say, "That's a good question. What do you think?" Freely acknowledge when you don't know the answer yourself, so your child learns its okay not to know. Ask, "What could we do to find out?" Model ways to find answers— going to the library, asking an expert, looking in the dictionary or encyclopedia, searching the Internet, conducting an experiment....

As we discussed in "How involved should you be?" on page 79, there's a tricky distinction between encouraging your child to engage in science, math, or anything else in a self-reliant way and doing too much for her and fostering a pattern of over-dependence. Open-ended questions are especially helpful in supporting your child's independent thinking.

Here are some ideas for questioning your child in a way that encourages learning and lets you see (without judgment) how he tends to think things through.

THE ART OF THE QUESTION

Sample questions that promote learning:

- What do you think happened?

- What do we know now?

- What do you mean by...?

- Why do you think that?

- Can you explain why?

- What might another explanation be?

- How is this different from (or the same as)...?

- What questions do you still have about this?

- How could we find out?

- How could we test our ideas?

A tip: Be sure to treat some "correct" answers with the same scrutiny as the incorrect ones, so that when you ask "What made you think that?" your child won't automatically assume his answer was wrong! And if it happens he is "wrong," a child will often—if asked this question and encouraged to look back over his work—find the error himself. He'll also be more inclined to assess his own work in other situations down the road.

Driving to School
by JB

Every morning for the last seven years, I've driven my children to school. Depending on the traffic on any given morning, or which way the car's parked, or whether we're late or not, or the changing preferences of the passengers, we choose different routes to drive to school.

"This way is shorter," pointed out my son one day. "How can you tell?" I asked. "Well it takes less time to get to school," he replied. "Do you think it's also a shorter distance?" I asked. He hesitated. "Let's figure it out."

And so began the months of investigation and data collection. We'd record our departure time, arrival time, and number of miles traveled (measured by the odometer). It turned out that the route that took the least amount of time was longer by several tenths of a mile than our second most frequently traveled route.

The questions just grew on their own from there. "Will it always take us the same amount of time if we go the same route?" "No, there's a range of time it takes." "Why does it take longer some days?" "Because of the traffic." "No, it has to do with when we leave." We tried going the same route several days in a row but leaving at different times. "See, I told you, it takes less time when we leave earlier." "But that's because there's more traffic later—it's because of the traffic!" We varied our route tremendously, choosing ways we'd never before thought to go. "Is it better to choose the routes that have more stoplights or more stop signs?" "Drive faster." "I'm driving as fast as I can on this route." "What do you mean?" "Different streets have different speed limits." "But the faster streets have more stop lights, so they aren't necessarily faster!"

We did our best to set up small controlled experiments. We'd go one way, one day; the next day we'd do everything the same, but change just one variable (the route, when we left, our average speed). Eventually the boys' interest waned, but by that time they'd been inculcated with a thorough knowledge of a path they'll be traveling for years more, and a good sense of how to investigate something they're curious about. There are now three children and three schools, but the route to investigating a question is imprinted for good.

Seize the Opportunities!

The science fair

*Genius is one-percent inspiration and
ninety-nine-percent perspiration.*
—Thomas Edison

By the time students reach middle school and high school they're increasingly able to define their own investigations and represent their inquiries in the form of experiments and projects. As their minds and abilities mature, their inquires become more complex. Coming up with interesting questions that are able to be investigated (a core element of scientific research); designing scientifically valid tests, experiments, or other procedures; making detailed observations and accurate measurements; using increasingly sophisticated methods for organizing and analyzing data—all these capabilities allow older students to dig deeper into the secrets of the natural world.

Science fairs present a great opportunity for you to support your child's inquiry. In the realm of a science- or math-fair project (or any other major project), knowing how much to help your child is an important consideration. (Remember "How involved should you be?") You might, as a start, consider these factors:

- The amount of support and preparation provided by your child's teacher;
- The role the teacher suggests you should play as a parent;

- Your child himself, and his confidence in science and in working independently.

Remember that the primary goal of a major project isn't the final product, but the process your child goes through in the course of the project. A few strategies to consider:

- Let your child take the lead and direct the project himself (especially as he gets older). It's best to step in yourself only if and when your child has exhausted his own capabilities. As in all activities with your child, showing him how to find out what he doesn't know is far more powerful than giving him the answer.
- Help him accept that not every question has an "answer"; science is still powerfully fueled by questions that date back to the dawn of the human brain!
- And, as always, nothing is more powerful—from earliest infancy to the most advanced old-age—than sincere encouragement.

For adolescents especially, pulled in so many directions by maturing bodies, social interactions, cultural influences, peer pressures, and a zillion other often conflicting interests, the awareness that hard work and clear focus can yield positive results is a powerful lesson. It's even more lasting and meaningful if it's learned in the course of work on something interesting, creative, compelling, and—best of all—something that was their own idea.

It all starts with **the question.** And thereby hangs a wonderful way to allow your child's natural curiosity to be given free rein. It's true, as the project actually gets planned for and underway, that the reins of practical reality, level of knowledge, time available, and procrastination tighten. But to start with, let the questions pour out!

Is your child having trouble coming up with a topic for her project? Can she find one she's truly curious about and interested in? Is she curious about why the cat rubs against people's legs or likes catnip? How the geese know which way to go when they migrate? How fast perfume spreads around the room? If no specific question comes to mind, maybe start by talking about the

---

Diary of a Successful Science Project
or
Are Pampers Damper?
by LB

Over the years of trying to be helpful with the various science projects of two wonderful daughters, what I now call "the lesson of the question" is one of the most important things I've learned. For a science project, the object is to arrive at a question that can be investigated. With this in mind, when it came time for my daughter Caitlin to choose a middle school science project, she discussed a wide variety of interests with me: plants, birds, music, cosmetics, our cats and dog—we ranged over quite a bit. Her past experience, including a plant-and-music experiment (allegedly show-ing that plants grow better to classical music and jazz than to hard rock) hadn't been so successful in controlling all the variables. She'd witnessed her older sister's experiments with culturing bacteria from the mouth and testing the effectiveness of different brands of mouthwash. That experi-ence, plus Caitlin's great love for little babies and her healthy skepticism when it comes to commercials, eventually led us to The Question: Which brand of diaper absorbs the most liquid; Luvs, Huggies, or Pampers?

Over the course of her project, Caitlin also described the substance in non-cloth diapers that helps them absorb liquid—Polysodium acrylate, or PSA. Researching this substance was a good way to show her that it's fine to do research in books or on the Internet to gain background for your project. She found out about the chemical structure. She discovered that a woman scientist, Mary Oliden Weaver, was one of the inventors of PSA. Caitlin included bar graphs, photographs of the (rather messy!) experi-mental procedure, and tables to document her scientific procedure. And the best part for me, she generated a list of additional questions that grew out of the project.

There were a number of excellent science-fair projects turned in at the middle school, indicating great creativity and ingenuity as well as some outstanding science teachers. A number of awards were offered at each grade level, for many different aspects, which was a nice way to do it. Caitlin's project, for example, won the award for "most clever title." But it also represented a couple of other important things—it was a project that allowed her to do things on her own, it had a result, and it offered some lessons about measurement criteria and how to do things better next time. (It also was a topic that lent itself to "toilet humor"....) In my daughter's recollection several years later, when I asked her which science project I should write about, she said this one. It turned out, she said, to be a lot of fun. And clearly, from my point of view, it was a memorable learning experience that will stay with her for years.

Oh, and which brand won the absorbency test? Here's what Caitlin wrote:

My data shows me that Luvs in fact absorbs the most liquid. Huggies came in second, while Pampers absorbs the least. Luvs absorbed 50 milliliters more than Huggies and 55 milliliters more than Pampers. We know this because we measured the absorbency according to the liquid buildup on top, deeper than a centimeter.

Further questions I might like to investigate are: which brand of diaper is the best deal, depending on both absorbency and cost; how do disposable diapers compare to cloth diapers; how much PSA is in each brand; and does PSA harm a baby's skin by causing rashes?

And in the way of all good scientists, Caitlin assessed her own work on the project:

Looking back at my experiment, there are a few things I could have done to improve it. I could have tested it more times, using more diapers. I also could have measured absorbency in many different ways. I also could have used cloth diapers as a control group, or in some other way made a comparison to cloth diapers. However, I do feel that I did a pretty good job on the experiment and I'm fairly confident of my findings.

Out of her very personal interests and experience, my daughter (like her sister before her) created a marvelous inquiry-driven science project that involved the whole family. And soon enough, next year's project will start with...a question....

(Pampers, Luvs, and Huggies are registered name-brand products.)

things she's most interested in, and how they might be related to science. See where that leads. The more motivated and interested your child is in the final topic the better—and the more likely she'll grapple with obstacles rather than giving up in frustration.

The kitchen table and beyond

The stories that follow may give you a feel for other kinds of inquiry opportunities that present themselves in everyday life. Many of the anecdotes elsewhere in the book are also strongly centered around inquiry—giving you an idea of how very natural it is!

"SCIENCE NIGHTS"—HOW ONE PARENT STEPPED IN

A parent with whom LHS staff spoke a few years ago tells a wonderful story. This father had noticed that his daughter, who'd always been excited by science before she began attending school, was now complaining that science was "boring" and "for boys."

Disturbed by the change in his daughter's attitude, he looked into her school's science program and concluded that it was woefully inadequate and not activity-based. This father, who had a business background but also a good grounding and real interest in education and science, decided to address the problem personally. He was determined to make science come alive for his daughter and some of her friends. Using a few science-experiment resource books and some basic materials, he created "Science Nights," conducting experiments with the children in his kitchen every Friday evening.

"The kids are enthralled," he says. "I've discovered that if they're just exposed to science in this way their natural curiosity and excitement can't be held down." He notes that these young students work harmoniously together to solve problems, and that both girls and boys participate fully. His daughter's friends now call to talk to him about science and ask what they're going to do next!

Rocks, Flags, Numbers, and Guessing Games

by John Erickson
Coordinator of Elementary Education Programs
Lawrence Hall of Science

With my daughter, who was about five years old, I was looking at the many rocks that decorated my aunt's garden.

"The greenish one is pretty," I said.

"This one?" she asked.

"The one I'm thinking of doesn't have stripes like that one." All at once, a comment on an interesting rock became a guessing game. A few more clues and she figured out which rock I'd been looking at. I suggested that she choose a rock and tell me about it so that I could be the one to guess. We played the game several more times. When we tried it with my younger children, they always wanted to look closely at the rock they chose—which made guessing very easy!—but it was still fun.

A few years ago we put up a world map in our kitchen. At the bottom it has pictures of the flags of all the countries. (Hammond publishes this map.) The game with the rocks is now the game with the flags. We not only have a fun activity in sorting and classifying, but we learn a little geography and current events too.

The children are now in upper elementary and middle school, but classifying flags is still a popular dinnertime sport. They like to throw in misleading clues. That's my fault. One day I said, "I'm looking at a flag that has the same amount of green as it has red on it." It was a while before they figured out that it had no green or red at all. Then one of them said, "I see a flag with a number of stars on it." Just by the way he said it I knew the number of stars was probably zero. I kept quiet while his sisters puzzled over the clues. One of my favorites was "I see a flag with one stripe." It was the flag of Libya, which is solid green—one wide green stripe going across, or maybe one really wide stripe going up and down.

My daughter, in middle school, was doing some math and gave me this question:

"I have a number that's a cubic number and it reads the same forward and backward."

We seemed to be playing the game again, but with numbers.

"Is it eight?" I asked.

"No. It's bigger. It's odd too."

That's the last clue I got, and I still haven't guessed.

The Ingredients of Inquiry
by JB

My children go on eating jags. They become crazy about a certain food for a month or two, and then it's on to the next great find. Currently it's Kung Pao chicken, a spicy Chinese dish of boned chicken pieces and peanuts served over rice. All three of them love it and will lobby for a dinner that includes it or beg to split a take-out order for an after-school snack.

But about "opportunity." One evening we decided to try four different orders of Kung Pao chicken from four different restaurants. Together our family brainstormed the key qualities, or properties, of great Kung Pao. We hit on five: 1) the chicken itself (tender and flavorful); 2) the sauce (not too gloppy); 3) the peanuts (crisp and fresh); 4) the vegetables (varied and fresh); and 5) the degree of spiciness (for this quality we each had different ideals!).

After the orders arrived, and as we tasted, we each rated every Kung Pao chicken order on a scale from one to 5 (5 was high)—first, for the merits of each individual quality, and at the end, for each restaurant's overall achievement of the five qualities. The scoring took considerable math in order to calculate average scores for each quality and an overall score for each version of the dish. It was fun to see how each of my children approached this Consumer Reports–like activity. My four-year-old gave either top or bottom ranks—never in the middle. My twelve-year-old was quite discerning. My nine-year-old marveled that he'd never paid attention to the sauce before.

Ranking a favorite dish was a lasting hit with my children. Now, whenever we taste a new Kung Pao chicken, there's lots of kibbitzing about how it ranks overall and which of the five qualities it excels in. I also see real evidence of how the process of isolating individual variables and using a quantitative evaluation system has influenced my children's outlook. I recently overheard my middle child and a friend rating their inventions, for instance. And my youngest child, the other day, protested about eating a requisite part of dinner. "It's not even a zero!"

It's not ONLY about creating future scientists

Inquiry-based learning doesn't hatch scientists. Well sometimes it does, but it also produces chefs and songwriters and orchidologists. Guided inquiry, by capitalizing on your child's natural curiosity and strengths, creates a "critical thinker." It guides a child to use all her senses, to explore her world with an insatiable "need to know." Inquiry-based learning is such a force in the education of a child that, like Dr. Glenn Seaborg, we have to use the word: it **empowers** her to perceive and question her world with a hungry confidence.

Valentine Business
by JB

They wanted to earn money, mostly. But my sons had also made some gorgeous watercolor paintings, about 4 x 6 each, as school valentines. I helped them duplicate the card-sized paintings on a color copier and mount them on construction paper. Everybody commented on how beautiful they looked. "Maybe we could start a valentine business and sell them. How much should we charge?" my two older sons asked. "How much do they cost to make?" I asked back.

And so started the first stage of the project. We agreed that the cost of selling the valentines would be the cost of duplicating them, buying construction paper, and buying envelopes. But figuring that out wasn't so simple. Saul, my oldest son, made a call to the closest copy center to learn the cost of a color copy. "What size copy?" asked the person at the copy center.

This sent us into a whole new discussion. "What size do we need?" asked Saul. "How many cards are you hoping to copy on each page?" I replied. We measured out three standard photocopy sizes on a large sheet of paper: letter-sized, legal-sized, and extra-large. The kids figured out that they could only fit four valentines on both the letter- and legal-sized papers, so why pay extra for a legal-sized color copy? But six valentines could fit on the extra-large size. "Let's choose the size that fits six!" suggested Leib, my middle son. "But how much does it cost compared to the letter-sized paper?" asked Saul, catching on. $1.50 for extra large; 60 cents for letter-sized. "Whoa! It costs more than twice as much but you can't even fit twice as many valentines!"

And so it continued. How much does construction paper cost? Well it depends where you buy it and how many sheets you buy. "How about envelopes?" Same thing. "Are you going to charge for your time?" asked their father. "What do you mean?" replied Leib. And so began a discussion of wage labor. They arrived on a price they wanted to charge per valentine. "Will you sell sets of valentines or individual valentines?" I asked. Big discussion. They arrived at 50 cents per valentine; $2.50 for a set of six. "How will you package the sets?" Fancy options were all rejected as the boys realized that it would add to their costs. Rubber bands were cheap. "How will you sell the valentines? Do you need a display? An order form?" Those cost, too. "We're going to have to find a cheaper way to make the valentines!" The largest cost by far was that of the color copy, so that seemed like the place where the savings should be focused. Saul phoned several other copy centers and found one with a cheaper price. "Yes!!"

But that was just the planning and the pricing! Many more problems arose that needed solving as the boys took orders, had to come up with a system for keeping track of orders, needed to make hundreds of valentines (and different numbers of each valentine, which made figuring out which combination of four valentines to copy on a page extremely challenging), and then arranged the delivery.

They learned to use and read a spreadsheet. They learned the marketing value of titling the valentines ("If you put the word 'love' in the title, then adults really like to buy it," concluded Leib.) They had to figure out how to increase their labor force. (Grandmothers will work for valentines.) And when they wanted, in a subsequent year, to make and sell valentines made by their new-to-the-enterprise, toddler-aged brother, they became engaged in an ethical discussion about who owned Jesse's artwork. "It's his artwork," agreed Leib. "Yeah, but we're doing all the work of making the valentines, and that's the hard part! Why should he get some of the money?!" protested Saul. (The solution was a modest one-time licensing fee paid to Jesse.)

My sons have sold valentines for three years now. Among all three boys, they have eight different sets of valentines. Their business has a name: atyourservice.com (though no Web site yet). Last year I showed them how to use a spreadsheet to project their expenses and examine their profit under different pricing plans. The lessons continue and abound with real-life math problems to solve.

They still mostly do it for the money.

IN CLOSING
Involvement Is
Its Own Reward

By the time most young people reach high school, they've already decided how they feel about math and science and whether they'll choose to pursue them further. This has a lot to do with whether they think they can "get" it. Their fears and reluctance must be caught and turned around long before then; and for that, your positive attitude and encouragement are critical. **Your involvement, even in small ways, is the single biggest determiner of your child's feelings about math and science for the rest of his life.**

By bringing your own sense of what's important to your child's education (with the help of this book and other resources), you can negotiate the issues of homework, involvement at school, newspaper stories about achievement, the direction of your child's education (and the nation's), and more.

As you get the hang of inquiry-based learning, you'll watch your child blossom, become engaged, and grow in self-confidence. Ideally, your child's questions will lead naturally to concepts you can help her grasp or that she can intuit on her own. As she discovers things for herself, there are ways in which you can encourage and focus the process. With your guidance, your child will learn how to inquire, draw conclusions, and construct better models of the world. In time, her logical thinking, problem-solving, and inquiry abilities will grow beyond the need for close guidance. She'll become an independent inquirer, able to understand and apply critical-thinking skills to many social and environmental issues. In gaining this self-reliance, she'll also come to collaborate effectively with others and experience the fulfillment of accomplishment, achievement, and—always—the thrill of "finding out more."

There's *no doubt* that in beginning or increasing your involvement in your child's education you're going to discover things about your child, yourself, the natural world, and education that repay you a hundredfold for your efforts. As our grandmothers said, "big oaks from little acorns grow." Even seemingly small encouraging words and little acts of parent involvement can have powerful and lasting positive effects. Like "learning" itself, involvement is its own exhilarating, cascading reward. We wish you all the joy it can offer.

The Making of a Scientist

BY RICHARD P. FEYNMAN, NOBEL LAUREATE IN PHYSICS

Before I was born, my father told my mother, "If it's a boy, he's going to be a scientist." When I was just a little kid, very small in a highchair, my father brought home a lot of little bathroom tiles—seconds—of different colors. We played with them, my father setting them up vertically on my highchair like dominoes, and I would push one end so they would all go down.

Then after a while, I'd help set them up. Pretty soon, we were setting them up in a more complicated way: two white tiles and a blue tile, two white tiles and a blue tile, and so on. When my mother saw that she said, "Leave the poor child alone. If he wants to put a blue tile, let him put a blue tile."

But my father said, "No, I want to show him what patterns are like and how interesting they are. It's a kind of elementary mathematics." So he started very early to tell me about the world and how interesting it is.

We had the *Encyclopedia Britannica* at home. When I was a small boy he used to sit me on his lap and read to me from the *Britannica*. We would be reading, say, about dinosaurs. It would be talking about the *Tyrannosaurus rex*, and it would say something like, "This dinosaur is twenty-five feet high and its head is six feet across."

My father would stop reading and say, "Now, let's see what that means. That would mean that if he stood in our front yard, he would be tall enough to put his head through our window up here." (We were on the second floor.) "But his head would be too wide to fit in the window." Everything he read to me he would translate as best he could into some reality.

It was very exciting and very, very interesting to think there were animals of such magnitude—and that they all died out, and that nobody knew why. I wasn't frightened that there would be one coming in my window as a consequence of this. But I learned from my father to translate: everything I read I try to figure out what it really means, what it's really saying.

We used to go to the Catskill Mountains, a place where people from New York City would go in the summer. The fathers would all return to New York to work during the week, and come back only for the weekend. On weekends, my father would take me for walks in the woods and he'd tell me about interesting things that were going on in the woods. When the other mothers saw this, they thought it was wonderful and that the other fathers should take their sons for walks. They tried to work on them but they didn't get anywhere at first. They wanted my father to take all the kids, but he didn't want to because he had a special relationship with me. So it ended up that the other fathers had to take their children for walks the next weekend.

The next Monday, when the fathers were all back at work, we kids were playing in a field. One kid says to me, "See that bird? What kind of bird is that?"

I said. "I haven't the slightest idea what kind of a bird it is."

He says, "It's a brown-throated thrush. Your father doesn't teach you anything!"

But it was the opposite. He had already taught me: "See that bird?" he says. "It's a Spencer's warbler." (I knew he didn't know the real name.) "Well, in Italian, it's a *Chutto Lapittida*. In Portuguese, it's a *Bom da Pieda*. In Chinese, it's a *Chung-long-tah*, and in Japanese, it's a *Katano Tekeda*. You can know the name of that bird in all the languages of the world, but when you're finished, you'll know absolutely nothing whatever about the bird. You'll only know about humans in different places, and what they call the bird. So let's look at the bird and see what it's *doing*—that's what counts." (I learned very early the difference between knowing the name of something and knowing something.)

He said, "For example, look: the bird pecks at its feathers all the time. See it walking around, pecking at its feathers?"

"Yeah."

He says, "Why do you think birds peck at their feathers?"

I said, "Well, maybe they mess up their feathers when they fly, so they're pecking them in order to straighten them out."

"All right," he says. "If that were the case, then they would peck a lot just after they've been flying. Then, after they've been on the ground a while, they wouldn't peck so much any more—you know what I mean?"

"Yeah."

He says, "Let's look and see if they peck more just after they land."

It wasn't hard to tell: there was not much difference between the birds that had been walking around a bit and those that had just landed. So I said, "I give up. Why does a bird peck at its feathers?"

"Because the lice are bothering it," he says. "The lice eat flakes of protein that come off its feathers."

He continued, "Each louse has some waxy stuff on its legs, and little mites eat that. The mites don't digest it perfectly, so they emit from their rear ends a sugar-like material, in which bacteria grow."

Finally he says "So you see, everywhere there's a source of food, there's *some* form of life that finds it."

Now, I knew that it may not have been exactly a louse, that it might not have been exactly true that a louse's legs have mites. That story was probably incorrect in *detail*, but what he was telling me was right in *principle*.

Another time, when I was older, he picked a leaf off of a tree. This leaf had a flaw, a thing we never look at much. The leaf was sort of deteriorated; it had a little brown line in the shape of a C, starting somewhere in the middle of the leaf and going out in a curl to the edge.

"Look at this brown line," he says. "It's narrow at the beginning and it's wider as it goes to the edge. What this is, is a fly—a blue fly with yellow eyes and green wings has come and laid an egg on this leaf. Then, when the egg hatches into a maggot (a caterpillar-like thing), it spends its whole life eating this leaf—that's where it gets its food. As it eats along, it leaves behind this brown trail of eaten leaf. As the maggot grows, the trail grows wider until he's grown to full size at the end of the leaf, where he turns into a fly—a blue fly with yellow eyes and green wings—who flies away and lays an egg on another leaf."

Again, I knew the details weren't precisely correct—it could have even been a beetle—but the idea that he was trying to explain to me was the amusing part of life: the whole thing is just reproduction. No matter how complicated the business is, the main point is to do it again!

Not having experiences with many fathers, I didn't realize how remarkable he was. How did he learn the deep principles of science and the love of it, what's behind it, and why it's worth doing? I never really asked him, because I just assumed that those were things that fathers knew.

My father taught me to notice things. One day, I was playing with an "express wagon," a little wagon with a railing around it. It had a ball in it, and when I pulled the wagon, I noticed something about the way the ball moved.

I went to my father and said, "Say, Pop, I noticed something. When I pull the wagon, the ball rolls to the back of the wagon. And when I'm pulling it along and I suddenly stop, the ball rolls to the front of the wagon. Why is that?"

"That, nobody knows," he said. "The general principal is that things which are moving tend to keep on moving, and things which are standing still tend to stand still, unless you push them hard. This tendency is called 'inertia,' but nobody knows why it's true." Now, that's a deep understanding. He didn't just give me the name.

He went on to say, "If you look from the side, you'll see that it's the back of the wagon that you're pulling against the ball, and the ball stands still. As a matter of fact, from the friction it starts to move forward a little bit in relation to the ground. It doesn't move back."

I ran back to the little wagon and set the ball up again and pulled the wagon. Looking sideways, I saw that indeed he was right. Relative to the sidewalk, it moved forward a little bit.

That's the way I was educated by my father, with those kinds of examples and discussions. It has motivated me for the rest of my life, and makes me interested in *all* the sciences. (It just happens I do physics better.)

I've been caught, so to speak—like someone who was given something wonderful when he was a child, and he's always looking for it again. I'm always looking, like a child, for the wonders I know I'm going to find— maybe not every time, but every once in a while.

APPENDIX A
Examples from National Education Standards

National Science Education Standards (NSES)

These standards outline what students need to know, be able to do, and understand to be scientifically literate at different grade levels. They're broken down in this way:

- Standards for science teaching
- Standards for professional development for teachers of science
- Standards for assessment in science education
- Standards for science content
- Standards for science education programs
- Standards for science education systems

The **science content** standards, perhaps likeliest to be of interest to you as a parent, outline what students should know, understand, and be able to do in the natural sciences over the course of K–12 education. They're divided into eight categories:

- **Unifying concepts and processes in science.** (These apply across all grade levels.)
- Science as inquiry. [K–4] [5–8] [9–12]
- Physical science. [K–4] [5–8] [9–12]
- Life science. [K–4] [5–8] [9–12]
- Earth and space science. [K–4] [5–8] [9–12]
- Science and technology. [K–4] [5–8] [9–12]
- Science in personal and social perspective. [K–4] [5–8] [9–12]
- History and nature of science. [K–4] [5–8] [9–12]

Here are excerpts from a couple of these categories, to give you a feel for what the *Standards* contain. The complete contents can be viewed (allow lots of time and a big cup of coffee) on the *NSES* Web site, www.nap.edu/html/nses/html.

EXCERPT

Science as Inquiry for Grades 5–8
CONTENT STANDARD A

As a result of activities in grades 5–8, all students should develop:

- ABILITIES NECESSARY TO DO SCIENTIFIC INQUIRY
- UNDERSTANDINGS ABOUT SCIENTIFIC INQUIRY

Fundamental abilities and concepts that underlie these standards:

ABILITIES NECESSARY TO DO SCIENTIFIC INQUIRY
- Identify questions that can be answered through scientific investigations....
- Design and conduct a scientific investigation....

- Use appropriate tools and techniques to gather, analyze, and interpret data....
- Develop descriptions, explanations, predictions, and models using evidence....
- Recognize and analyze alternative explanations and predictions....
- Communicate scientific procedures and explanations....

UNDERSTANDINGS ABOUT SCIENTIFIC INQUIRY
- Different kinds of questions suggest different kinds of scientific investigations....
- Current scientific knowledge and understanding guide scientific investigations....
- Mathematics is important in all aspects of scientific inquiry.
- Technology used to gather data enhances accuracy and allows scientists to analyze and quantify results of investigations.
- Scientific explanations emphasize evidence, have logically consistent arguments, and use scientific principles, models, and theories.
- Science advances through legitimate skepticism....
- Scientific investigations sometimes result in new ideas and phenomena for study, generate new methods or procedures for an investigation, or develop new technologies to improve the collection of data....

EXCERPT

Earth and Space Science for Grades K–4
CONTENT STANDARD D

As a result of their activities in grades K–4, all students should develop an understanding of:

- PROPERTIES OF EARTH MATERIALS
- OBJECTS IN THE SKY
- CHANGES IN EARTH AND SKY

Fundamental concepts and principles that underlie these standards:

PROPERTIES OF EARTH MATERIALS
- Earth materials are solid rocks and soils, water, and the gases of the atmosphere....
- Soils have properties of color and texture, capacity to retain water, and ability to support the growth of many kinds of plants, including those in our food supply.
- Fossils provide evidence about the plants and animals that lived long ago and the nature of the environment at that time.

OBJECTS IN THE SKY
- The sun, moon, stars, clouds, birds, and airplanes all have properties, locations, and movements that can be observed and described.
- The sun provides the light and heat necessary to maintain the temperature of the earth.

CHANGES IN THE EARTH AND SKY
- The surface of the earth changes....
- Weather changes from day to day and over the seasons....
- Objects in the sky have patterns of movement....

Benchmarks for Science Literacy (BSL)

Benchmarks is a publication of Project 2061, a curriculum-reform project that promotes literacy in science, mathematics, and technology "in order to help people live interesting, responsible, and productive lives." *Benchmarks* specifies how students should progress toward science literacy by recommending what they should know and be able to do in these subjects by the end of grades 2, 5, 8, and 12. (An earlier publication by Project 2061, *Science for All Americans*, addressed science literacy after high school.) The document is arranged by subject:

• The Nature of Science	• Human Society
• The Nature of Mathematics	• The Designed World
• The Nature of Technology	• The Mathematical World
• The Physical Setting	• Historical Perspectives
• The Living Environment	• Common Themes
• The Human Organism	• Habits of Mind

Picking one of these subjects at random—**Nature of Science**—let's look at how it's divided:

A. The Scientific World View
Kindergarten–Grade 2
Grades 3–5
Grades 6–8
Grades 9–12
B. Scientific Inquiry
Kindergarten–Grade 2
Grades 3–5
Grades 6–8
Grades 9–12
C. The Scientific Enterprise
Kindergarten–Grade 2
Grades 3–5
Grades 6–8
Grades 9–12

Here are excerpts from a couple of these categories:

EXCERPT

A. The Scientific World View: Grades 9–12

By the end of the 12th grade, students should know that:
- Scientists assume that the universe is a vast single system in which the

basic rules are the same everywhere. The rules may range from very simple to extremely complex, but scientists operate on the belief that the rules can be discovered by careful, systematic study.

- From time to time, major shifts occur in the scientific view of how the world works. More often, however, the changes that take place in the body of scientific knowledge are small modifications of prior knowledge. Change and continuity are persistent features of science.

- No matter how well one theory fits observations, a new theory might fit them just as well or better, or might fit a wider range of observations. In science, the testing, revising, and occasional discarding of theories, new and old, never ends. This ongoing process leads to an increasingly better understanding of how things work in the world, but not to absolute truth. Evidence for the value of this approach is given by the improving ability of scientists to offer reliable explanations and make accurate predictions.

EXCERPT

C. The Scientific Enterprise: Grades 3–5

By the end of the 5th grade, students should know that:

- Science is an adventure that people everywhere can take part in, as they have for many centuries.
- Clear communication is an essential part of doing science. It enables scientists to inform others about their work, expose their ideas to criticism by other scientists, and stay informed about scientific discoveries around the world.
- Doing science involves many different kinds of work and engages men and women of all ages and backgrounds.

Principles and Standards for School Mathematics

Principles and Standards is the most recent document produced by the National Council of Teachers of Mathematics (NCTM) to address math-education reform. Previous publications, which you may recognize, were *Curriculum and Evaluation Standards for School Mathematics* (1989), *Professional Standards for Teaching Mathematics* (1991), and *Assessment Standards for School Mathematics* (1995).

Principles and Standards emphasizes the need for mathematical literacy, and proposes what's possible for students in grades preK–12 to learn about certain math content areas at certain levels and under certain teaching conditions. The document is organized into eight chapters:

1. A Vision for School Mathematics
2. Principles for School Mathematics
3. Standards for School Mathematics
4. Standards for Grades Pre-K–2
5. Standards for Grades 3–5
6. Standards for Grades 6–8
7. Standards for Grades 9–12
8. Working Together to Achieve the Vision

The **Principles for School Mathematics,** to again take one chapter as an example, don't refer to specific mathematics content or processes, as standards do, but to what the document calls "critical issues deeply intertwined with school mathematics programs." The six principles are:

1. **Equity.** Excellence in mathematics education requires equity—high expectations and strong support for all students.
2. **Curriculum.** A curriculum is more than a collection of activities: It must be coherent, focused on important mathematics, and well articulated across the grades.
3. **Teaching.** Effective mathematics teaching requires understanding what students know and need to learn and then challenging and supporting them to learn it well.
4. **Learning.** Students must learn mathematics with understanding, actively building new knowledge from experience and prior knowledge.
5. **Assessment.** Assessment should support the learning of important mathematics and furnish useful information to both teachers and students.
6. **Technology.** Technology is essential in teaching and learning mathematics; it influences the mathematics that is taught and enhances students' learning.

Let's take the **Equity Principle,** for instance, a step further. Here are the kinds of ideas *Principles and Standards* describes, with brief excerpts from the lengthier explanations:

- **Equity requires high expectations and worthwhile opportunities for all.**
  The vision of equity in mathematics education challenges a pervasive societal belief in North America that only some students are capable of learning mathematics. This belief…leads to low expectations for too many students. Low expectations are especially problematic because students who live in poverty, students who are not native speakers of English, students with disabilities, females, and many nonwhite students have traditionally been far more likely than their counterparts in other demographic groups to be the victims of low expectations. Expectations must be raised—mathematics can and must be learned by *all* students.…
- **Equity requires accommodating differences to help everyone learn mathematics.**
  Higher expectations are necessary, but they are not sufficient to accomplish the goal of an equitable school mathematics education for all students. All students should have access to an excellent and equitable mathematics program that provides solid support for their learning and is responsive to their prior knowledge, intellectual strengths, and personal interests.…
- **Equity requires resources and support for all classrooms and all students.**
  Well-documented examples demonstrate that all children, including those who have been traditionally underserved, can learn mathematics when they have access to high-quality instructional programs that support their learning.… These examples should become the norm rather than the exception in school mathematics education.…

Standards for Grades 6–8, the sixth chapter in *Principles and Standards*, states that middle-grade students should see mathematics "as an exciting, useful, and creative field of study." These are the standards it proposes:

Number and Operations Standard for Grades 6–8

Instructional programs from prekindergarten through grade 12 should enable all students to:

1. UNDERSTAND NUMBERS, WAYS OF REPRESENTING NUMBERS, RELATIONSHIPS AMONG NUMBERS, AND NUMBER SYSTEMS

In grades 6–8 all students should:
- work flexibly with fractions, decimals, and percents to solve problems;
- compare and order fractions, decimals, and percents efficiently and find their approximate locations on a number line;
- develop meaning for percents greater than 100 and less than 1;
- understand and use ratios and proportions to represent quantitative relationships;
- develop an understanding of large numbers and recognize and appropriately use exponential, scientific, and calculator notation;
- use factors, multiples, prime factorization, and relatively prime numbers to solve problems;
- develop meaning for integers and represent and compare quantities with them.

2. UNDERSTAND MEANINGS OF OPERATIONS AND HOW THEY RELATE TO ONE ANOTHER

In grades 6–8 all students should:
- understand the meaning and effects of arithmetic operations with fractions, decimals, and integers;
- use the associative and commutative properties of addition and multiplication and the distributive property of multiplication over addition to simplify computations with integers, fractions, and decimals;
- understand and use the inverse relationships of addition and subtraction, multiplication and division, and squaring and finding square roots to simplify computations and solve problems.

3. COMPUTE FLUENTLY AND MAKE REASONABLE ESTIMATES

In grades 6–8 all students should:

- select appropriate methods and tools for computing with fractions and decimals from among mental computation, estimation, calculators or computers, and paper and pencil, depending on the situation, and apply the selected methods;
- develop and analyze algorithms for computing with fractions, decimals, and integers and develop fluency in their use;
- develop and use strategies to estimate the results of rational-number computations and judge the reasonableness of the results;
- develop, analyze, and explain methods for solving problems involving proportions, such as scaling and finding equivalent ratios.

APPENDIX B
Recommended Family Resources

GEMS® Home Science Kits from Scientific Explorer

The Great Explorations in Math and Science (GEMS) program of the Lawrence Hall of Science has partnered with Scientific Explorer to create home science kits for children and their families. Based on some of the most popular activities developed at LHS, each kit has been designed, tested, and refined by GEMS staff for maximum fun and learning potential. The goal is to capture the educational excitement for which LHS is famous and make it available to families, at home. The activities develop inquiry skills and build positive attitudes toward science and learning.

Soda Pop Science, Mystery Detective, and **Messages from Space** kicked off the series, and kits are added every year.

Available at toy and educational stores throughout the country,
or from:
Lawrence Hall of Science
University of California
Berkeley, CA 94720-5200
(510) 642-1016
www.lhsgems.org/gemsathome.html

FAMILY MATH

FAMILY MATH, one of the EQUALS programs out of the Lawrence Hall of Science, is designed to promote math equity and access to all students, regardless of race, gender, or class. The goal of the program is to promote parents as effective partners in helping their children succeed in mathematics. It focuses entirely on diverse families learning mathematics together, using inexpensive materials and hands-on, problem-solving activities. Children and adults come together once a week for several weeks, doing most activities in small groups with two or three families working together. Best of all, they put into practice at home the ideas they've developed in class.

FAMILY MATH activities illustrate concepts that reinforce the school math curriculum and demonstrate the importance of mathematics to future work and education. The program transforms student attitudes toward mathematics, gives parents a sense of confidence, reenergizes teachers, and builds community.

FAMILY MATH
Lawrence Hall of Science
University of California
Berkeley, CA 94720-5200
(510) 643-6525 or (510) 643-6350
http://www.lhs.berkeley.edu/equals

THE ALGEBRA PROJECT

The Algebra Project was born in the 1980s out of one parent's concern with the mathematics education of his children in the public schools of Cambridge, Massachusetts. The parent was renowned 1960s civil-rights activist Robert P. Moses, also a math educator who'd taught in New York and Tanzania. The main goal of the Algebra Project is to assist students in inner-city and rural areas achieve mathematics literacy. The Algebra Project has developed curricular materials, trained teachers and trainers of teachers, provided ongoing professional-development support, and developed community-involvement activities for schools seeking to achieve a systemic change in mathematics education.

The Algebra Project reaches approximately 10,000 students and 300 teachers a year in 10 states and 28 localities, with a particular focus on the southern U.S. The project has demonstrably improved student performance and increased the numbers of students enrolling in college prep math classes. Graduates of the Algebra Project have also formed a "Young Peoples' Project" that recruits, trains, and deploys high school and college-age youth to work with younger children in a variety of math learning contexts.

The Algebra Project, Inc.
99 Bishop Allen Drive
Cambridge, MA 02139
(617) 491-0200 ext.130
www.algebra.org

Parent Portal—a Web Site for Parents
www.lhsparent.org

Like this book (and in fact designed to be a companion site), Parent Portal is a highly accessible, practical Web resource for adult caregivers of school-aged children. Created to provide all parents with clear and practical advice, active "involvement" tools, and solid math and science resources for helping children learn, Parent Portal is deliberately jargon-free and practical: a one-of-a-kind resource for every hands-on parent. It includes sections on supporting your child's schooling at home; research findings about how parent involvement promotes academic success; great things to do at home with math and science; finding out what your child knows and how she learns best; and a "communication corner" for sharing insights and advice.

Active Parenting® Publishers

Founded by noted family therapist Michael H. Popkin and awarded the Positive Parenting Award from the National Council for Children's Rights, this organization produces innovative, research-based education resources for parents, children, and teachers. APP published the excellent book *Helping Your Child Succeed in School,* and in 1983 created the Active Parenting Discussion Program (APDP), the world's first video-based program directed at parent education. The series includes *Active Parenting Today, Active Parenting of Teens, Parenting Children Ages 1–4, Cooperative Parenting and Divorce,* and *1,2,3,4, Parents!,* among others. (Several are also available in Spanish.) APP also focuses on current issues such as drug abuse and school violence, and has formed alliances with other leading producers of parenting education products. APP offerings are widely distributed in the United States and in Canada, Japan, Korea, and Sweden.

Active Parenting Publishers
810-B Franklin Ct.
Marietta, GA 30067
(800) 825-0060
www.activeparenting.com

MARE Field Trip Leaders' Guide
Marine Activities, Resources & Education (MARE)

The MARE program at LHS has created field-trip resources for teachers that are packed with advice and organizational tips easily adaptable by parents. The MARE guides provide accessible information for planning simple interpretive field trips with your children. Each opens with a primer on how to interact with nature, including solid suggestions for how to use all our senses, encourage inquiry, make careful observations, and be good stewards of the natural places we visit. The guides include checklists, equipment suggestions, and ideas for making your trips educationally meaningful by planning conceptual goals. They include sample activities to focus and inspire your child no matter what your objectives for the field trips. Guides are available for kelp forests, sandy beaches, rocky seashores, and field-trip sites in Central California. Each environment is described in detail, and suggestions can be adapted to other locations.

Marine Activities, Resources & Education (MARE)
Lawrence Hall of Science
University of California
Berkeley, CA 94720-5200
(510) 642-5008
www.lhs.berkeley.edu:80/MARE/

FAMILY HEALTH

FAMILY HEALTH is a collaborative project of LHS and the UC Berkeley School of Public Health, Center for Community Wellness. The program develops and disseminates exciting health-science activities and follow-up materials, and works with parents, teens, teachers, and health professionals to build partnerships within the community through conferences, workshops, and family events.

FAMILY HEALTH's materials and activities on nutrition, heart fitness, and lifestyle issues are interactive and can easily be taught by parents. Activities can also be presented as a school-wide program to teach children and families science principles and decision-making skills that promote healthy choices and activities. Other components include theater performances in which secondary school students present skits and demonstrations on health issues; a Medical Mystery Festival, which takes children and adults through engaging activities that simulate medical investigations; and the *Wellness Guide*, a publication that highlights health information and resources (also available in a Spanish version, *La guia del bienestar*).

FAMILY HEALTH
Lawrence Hall of Science
University of California
Berkeley, CA 94720-5200
(510) 643-5537
www.lhs.berkeley.edu/familyhealth/

The PRIMES Project
Parents Rediscovering and Interacting with Math and Engaging Schools

Funded by the National Science Foundation, the PRIMES Project was created in 1997 to build the confidence of parents who feel anxious about math and worried about helping their children in the subject. PRIMES shows parents that they already use math fluently in everyday life and can be a real support to their middle school children. Using workshops, the parent resource guide *Middle School Math: What Every Parent Should Know*, and a video called *The Family Angle*, the PRIMES Project shows parents how they already use math every day—from setting family budgets to discussing baseball batting averages—and provides insight into innovative teaching methods schools are beginning to adopt. PRIMES's strong message is that parents are the most valuable advocates children have when they're learning math.

The PRIMES Project
Stanford University
School of Education
Cubberley Building
485 Lasuen Mall
Stanford, CA 94305
(650) 723-8422

Parent Partners

Parent Partners: Workshops to Foster School/Home/Family Partnerships is a publication of the Great Explorations in Math and Science (GEMS) program of the Lawrence Hall of Science. It provides activist parents, community leaders, and teachers with concrete, practical ways to help children succeed in school. *Parent Partners* provides interactive ways for parents to become better informed about current educational issues. While there's a strong focus on science and math, many of the lessons apply to education in general. The book includes:

- Step-by-step presentation instructions for three parent-education sessions;
- Numerous clear, jargon-free handouts that convey key information for parents and other adult caregivers; and
- Short, engaging digests of research findings that can motivate parents to get involved.

Parent Partners (and the Spanish-language version, *Padres participantes*) is available at many bookstores or directly from GEMS.

www.lhsgems.org/gemparpar.html
or see the GEMS contact information below

Great Explorations in Math and Science—GEMS®

GEMS is widely recognized as a leading educational resource for the advancement of inquiry-based science and math. (The Expert Panel on Mathematics and Science Education recognized GEMS for educational excellence in 2001.) GEMS is the workplace of the authors of this book.

At least 10 million students have experienced GEMS activities to date. The program includes teacher's guides for preK–8, classroom materials kits, handbooks on special educational topics, professional development for teachers, innovative parent-education workshops, home science kits, and a growing international network of GEMS centers and sites.

Many parents have used GEMS units with great success; the guides trigger ideas for your child's science projects, support home-school science curricula, provide hands-on activities at community gatherings, and are a great way to interest your child's teacher in inquiry-based math and science. GEMS activities use readily available materials and require no special training in math or science.

Great Explorations in Math and Science (GEMS)
Lawrence Hall of Science
University of California
Berkeley, CA 94720-5200
(510) 642-7771
www.lhsgems.org

APPENDIX C

Anatomy of a Rubric—1

The Science Education for Public Understanding Program (SEPUP) at the Lawrence Hall of Science is an issues-based science curriculum. SEPUP uses an assessment system developed with the University of California's Berkeley Evaluation & Assessment Research (BEAR) Center to help students, teachers, and administrators gain essential insight into what students are learning over their yearlong course. At the heart of the SEPUP/BEAR assessment system is the concept of progress variables, concepts or abilities on which students are scored. Here are the five main variables:

1. Designing and Conducting Investigations—designing a scientific experiment, performing laboratory procedures to collect data, recording and organizing data, and analyzing and interpreting the results of an experiment.

2. Evidence and Tradeoffs—identifying objective scientific evidence as well as evaluating the advantages and disadvantages of different possible solutions to a problem based on the available evidence. *(See the rubric for this variable in the table, opposite.)*

3. Understanding Concepts—understanding scientific concepts (such as properties and interactions of materials, energy, or thresholds) in order to apply the relevant scientific concepts to the solution of problems.

4. Communicating Scientific Information—organizing and presenting results of an experiment, or explaining the process of gathering evidence and weighing tradeoffs in selecting a solution to a problem, effectively, and free of technical errors.

5. Group Interaction—developing skill in collaborating with teammates to complete a task (such as a lab experiment), sharing the work of the activity, and contributing ideas to generate solutions to a given problem.

Here's a SEPUP assessment assignment, based on activities, readings, reflection, and discussion students have engaged in:

You are a public health official who works in the Water Department. Your supervisor has asked you to respond to the public's concern about water chlorination at the next City Council meeting. Prepare a written response explaining the issues raised in the newspaper articles. Be sure to discuss the advantages and disadvantages of chlorinated drinking water in your response, and then explain your recommendation about whether the water should be chlorinated.

This assignment would be scored to assess student achievement in several progress variables. The rubric (scoring guide) for the "Evidence and Tradeoffs" variable in this assignment is summarized in the following table.

SCORE	USING EVIDENCE	USING EVIDENCE TO MAKE TRADEOFFS
	Response uses objective reason(s) based on relevant evidence to argue for or against an option.	**Response recognizes multiple perspectives of issue and explains each perspective using objective reasons, supported by evidence, in order to make choice.**
4	Accomplishes Score Level 3 [below] AND goes beyond in some significant way, such as questioning or justifying the source, validity, and/or quantity of evidence.	Accomplishes Score Level 3 [below] AND goes beyond in some significant way, such as suggesting additional evidence beyond the activity that would further influence choices in specific ways, OR questioning the source, validity, and/or quantity of evidence and explaining how it influences choice.
3	Provides major objective reasons AND supports each with relevant and accurate evidence.	Makes a tradeoff using at least two perspectives or options AND provides objective reasons, supported by relevant and accurate evidence, for each option.
2	Provides some objective reasons AND some supporting evidence, BUT at least one reason is missing and/or part of the evidence is incomplete.	States at least two perspectives or options AND provides some objective reasons using some relevant evidence BUT reasons are incomplete and/or part of the evidence is missing; OR only one complete and accurate option has been provided.
1	Provides only subjective reasons (opinions) for choice and/or uses inaccurate or irrelevant evidence from the activity.	States at least one perspective/option BUT only provides subjective reasons (opinions) and/or uses inaccurate or irrelevant evidence.
0	Response missing, illegible or lacks reasons AND offers no evidence to support decision made.	Response missing, illegible or lacks reasons AND offers no evidence to support decision made.
X	Student had no opportunity to respond.	Student had no opportunity to respond.

Anatomy of a Rubric—2

This rubric, adapted from the GEMS guide Dry Ice Investigations, *sets out the criteria for a scientific investigation. It makes a distinction between a scientific experiment and a systematic observation. Students can choose either one, as both are very important in scientific work. The activities and discussions in the unit help students understand what each one is and when it makes sense to do an experiment and when a systematic observation. Here are questions a teacher might use to score a student's work:*

1. The student selected a question that can be investigated. The question:
   - ☐ uses available materials and equipment
   - ☐ is safe and realistic
   - ☐ can be answered with a single investigation (in other words, it's not too big a question)
   - ☐ is a "measuring" question, a "what happens if" question, or a "comparison" question, rather than a question that asks only a broad "how" or "why"

2. The student selected an appropriate kind of investigation. The student:
   - ☐ decided to do a *systematic observation,* because a "measuring" or "what happens if" question was chosen

     or
   - ☐ decided to do an *experiment,* because a "comparison" question was chosen or created out of another question

3. The student designed the investigation well.
   a) For systematic observations, the student:
   - ☐ planned the conditions (variables)
   - ☐ identified possible outcome variables
   - ☐ used clear and careful procedure that takes variables into account

   b) For experiments, the student:
   - ☐ identified test variable
   - ☐ controlled the variables
   - ☐ identified outcome variables
   - ☐ used clear and careful procedure that takes variables into account

4. The student used careful reasoning. The student:
   - ☐ used data and results to support conclusions
   - ☐ suggested a well-reasoned explanation
   - ☐ thought through problems and additional questions
   - ☐ clearly expressed ideas through writing and diagrams so others can understand the investigation and the reasoning.

Resources & Recommended Reading

Beating the Odds
Raising Academically Successful African American Males
Freeman A. Hrabowski III, Kenneth I. Maton, and Geoffrey L. Greif
Oxford University Press, 1998

A compelling switch from the often gloomy coverage of the academic achievement of African Americans. Many concrete examples and suggestions directly from parents and their high-achieving children.

Helping Your Child Succeed in School
A Guide for Parents of 4 to 14 Year Olds
Michael H. Popkin, Bettie B. Youngs, Jane M. Healy
Active Parenting Publishers, 1995

Solid suggestions that concur with many of the ideas expressed in this book.

Motivated Minds
Raising Children to Love Learning
Deborah Stipek and Kathy Seal
Henry Holt and Company, 2001

Confidence-inspiring with a minimum of touchy-feely. Good ideas and the research-based reasoning behind them.

Parent Power
90 Winning Ways to Be Involved and Help Your Child Get the Most Out of School
Roberta Kirshbaum with Robin Dellabough
Hyperion, 1998

Short, digestible suggestions that run the gamut from simple to highly ambitious. A good addition to your night-stand stack.

EDUCATIONAL REFORM

A Nation at Risk
The Imperative for Educational Reform
National Commission on Excellence in Education, 1983

A sobering heads-up—even two decades ago—about the consequences of educational inaction, with tangible solutions.

A New Generation of Evidence
The Family Is Critical to Student Achievement
Anne T. Henderson and Nancy Berla
Center for Law and Education, 1994

An anthology of keenly researched examples of how, and how profoundly, families influence children's academic success.

Before It's Too Late
A Report to the Nation from the National Commission on Mathematics and Science Teaching
National Commission on Mathematics and Science Teaching for the 21st Century, 2000

Like *A Nation at Risk*, a call to action in mathematics- and science-education reform, and a picture of the impact on society—for better or for much, much worse.

Every Child a Scientist: Achieving Scientific Literacy for All
How to Use the *National Science Education Standards* to Improve Your Child's School Science Program
National Research Council
National Academy Press, 1998

A slender overview of the state of science education today, the components of reform, and a vision for quality.

Falling Through the Net: Defining the Digital Divide
A Report on the Telecommunications and Information Technology Gap in America
U.S. Department of Commerce, 2000
Also available online: www.ntia.doc.gov/ntiahome/digitaldivide/

A revealing assessment of computer and Internet use across the country, and how it breaks down by ethnicity and socioeconomic lines.

High Stakes
Testing for Tracking, Promotion, and Graduation
National Research Council, 1999

Another quality NRC contribution to understanding the state of educational reform today; in this case, the politics and social consequences of testing.

The New Meaning of Educational Change
Michael G. Fullan with Suzanne Stiegelbauer
Teachers College Press, 2001

Fullan explains the frailties and strengths of educational reform, and addresses hard strategies for the future.

SCHOOLS AND SCHOOL REFORM

At Home in Our Schools
A Guide to Schoolwide Activities That Build Community
Child Development Project Development Studies Center, 1994

School-wide events that partner parents, schools, and the community .

Beyond the Classroom
Why School Reform Has Failed and What Parents Need to Do
Laurence Steinberg, with B. Bradford Brown and Sanford M. Dornbusch
Simon and Schuster/Touchstone, 1997

A navigable mix of research and commentary, particularly engrossing on the subjects of student engagement and family influence.

Parent Partners
Workshops to Foster School/Home/Family Partnerships
Jacqueline Barber with Lynn Barakos and Lincoln Bergman
GEMS, 2000

Blueprints for teacher-parent workshops to motivate caregivers of school-aged children to become actively involved in their children's education. Many tools and ideas from *Parent Partners* went into this book.

Savage Inequalities
Children in America's Schools
Jonathan Kozol
HarperPerennial, 1992

An impassioned indictment of how our public school system fails—and undermines—children of poverty.

Science for All Children
A Guide to Improving Elementary Science Education in Your School District
National Science Resources Center
National Academy Press, 1997

An instructive description of inquiry-based science, its building blocks, and sites where it's been implemented with great success.

Shaking Up the Schoolhouse
How to Support and Sustain Educational Innovation
Phillip C. Schlechty
Jossey-Bass, 2000

Written for school personnel, but a great window into the planning and foundation building that should be part of every school community's long-term vision.

The Schools Our Children Deserve
Moving Beyond Traditional Classrooms and "Tougher Standards"
Alfie Kohn
Houghton Mifflin, 1999

This veteran educator swipes at the grades-obsessed back-to-basics movement, arguing persuasively for schooling that reflects—and prepares students for—a more well-rounded world.

What's Worth Fighting for in the Principalship
Michael G. Fullan
Teachers College Press, 1997

What's Worth Fighting for in Your School
Michael G. Fullan, Andy Hargreaves
Teachers College Press, 1996

What's Worth Fighting for Out There
Andy Hargreaves, Michael G. Fullan
Teachers College Press, 1998

While geared toward educators and administrators, this trilogy offers real insight into the makings of successful school change at several levels. The last book includes a brief section specifically for parents.

HOMEWORK

How to Help Your Child with Homework
Every caring parent's guide to encouraging good study habits and ending the homework wars (for parents of children ages 6–13)
Marguerite C. Radencich and Jeanne Shay Schumm
Free Spirit Publishing, 1997

A conventional how-to book, covering the fundamentals as well as a few inspired ideas.

101 Educational Conversations You Should Have with Your Child
Vito Perrone
Chelsea House, 1995

This series offers modest but usable ideas on how to open channels with your child and support her schooling at home.

"School Success Begins at Home"
Tom Loveless
USA Weekend, August 24-25, 2001

One side of the coin; a vehement argument in favor of homework.

The End of Homework
How Homework Disrupts Families, Overburdens Children, and Limits Learning
Etta Kralovec and John Buell
Beacon Press, 2000

The other side of the story; the strong case for moderation and attention to our children's saturation points.

THE BRAIN AND HOW WE LEARN

How People Learn
Brain, Mind, Experience, and School
National Research Council
National Academy Press, 2000

Sometimes heavy going, but attentive to the relationship between learning behavior and what happens in the classroom.

Magic Trees of the Mind
How to Nurture Your Child's Intelligence, Creativity, and Healthy Emotions from Birth Through Adolescence
Marian Diamond and Janet Hopson
Penguin Books, 1999

An absorbing, anecdote-filled look at how the mind works and how to lay the groundwork for your child's learning at every stage of her development.

The Scientist in the Crib
What Early Learning Tells Us About the Mind
Alison Gopnik, Andrew N. Meltzoff and Patricia K. Kuhl
HarperPerennial, 2000

Delightful and intriguing; a readable mix of "Aha!" moments and hard science.

GENDER DIFFERENCES

Brain Sex
The Real Difference Between Men & Women
Anne Moir and David Jessel
Dell Books, reissue ed. 1993

As the brains of infant girls and boys go their separate structural and functional ways, how male is male and what does female mean?

Boys and Girls Learn Differently!
A Guide for Teachers and Parents
Michael Gurian and Patricia Henley with Terry Trueman

Jossey-Bass, 2001

Turns the lens on subtle and not-so-subtle distinctions between girls' and boys' learning styles and how those differences affect the classroom. The authors include proposed strategies for leveling the field.

Real Boys
Rescuing Our Sons from the Myths of Boyhood
William S. Pollack
Henry Holt/Owl Books, 1999

Real Boys' Voices
William S. Pollack with Todd Shuster
Penguin Books, 2001

Real Boys Workbook
The Definitive Guide to Understanding and Interacting with Boys of All Ages
William S. Pollack and Kathleen Cushman
Villard Books, 2001

These three books unveil and reexamine our conventional understandings of what it means to be male in the modern world.

MULTIPLE INTELLIGENCES

Building on Children's Strengths
The Experience of Project Spectrum
Jie-Qi Chen, Mara Krechevsky, and Julie Viens with Emily Isberg
Teachers College Press, 1998

Not as lay-reader-friendly as it promises, but full of examples of how the companion theories of "multiple intelligences" and "nonuniversal development" are used in real settings.

Frames of Mind
The Theory of Multiple Intelligences
Howard Gardner
Basic Books, 10th-anniversary ed. 1993

The classic. A readable introduction to the idea that we learn and excel in many ways, based on a variety of individual strengths.

In Their Own Way
Discovering and Encouraging Your Child's Personal Learning Style
Thomas Armstrong
J P Tarcher, 2000

Perhaps the most popular multiple-intelligences resource available; intuitive and full of common sense.

ABOUT STANDARDS

The following publications offer a range of critiques of standards-based reform and "teaching to the test."

"Everything you learned in kindergarten is no longer enough; California standards create high expectations for little kids"
Katherine Seligman
San Francisco Chronicle, August 26, 2001
Also available online via www.sfgate.com

One Size Fits Few
The Folly of Educational Standards
Susan Ohanian
Heinemann, 1999

Standardized Minds
The High Price of America's Testing Culture and What We Can Do to Change It
Peter Sacks
Perseus Books, 2001

"The Mismeasure of Learning"
Lauren Resnick
Education Next, Fall 2001
Also available online via www.educationnext.org

"The Test Mess"
James Traub

The New York Times, April 7, 2002
Also available online via www.nytimes.com (search for author; registration is free)

Will Standards Save Public Education?
Deborah Meier, Jonathan Kozol, Joshua Cohen
Beacon Press, 2000

WHERE TO READ THE STANDARDS

Benchmarks for Science Literacy
American Association for the Advancement of Science (AAAS) Project 2061
Oxford University Press, 1993
online: www.project2061.org/tools/benchol/bolframe.htm

National Science Education Standards
National Research Council
National Academy Press, 1996
online: www.nap.edu/books/0309053269/html

Principles and Standards for School Mathematics
National Council of Teachers of Mathematics (NCTM), 2000
online: http://standards.nctm.org

ASSESSMENT

Doing What Scientists Do
Children Learn to Investigate Their World
Ellen Doris
Heinemann, 1991

Though this book is written for elementary school teachers, the chapter called "Interpreting Children's Work" is a fascinating look at the kinds of things any parent might look for when trying to learn what her child understands.

Knowing What Students Know
The Science and Design of Educational Assessment
National Research Council
National Academy Press, 2001

Not easy reading, but a thorough look at the reasons for and kinds of assessment tools used in schools; their design, their audiences, and their relative benefits.

The Teaching of Science in Primary Schools
Wynne Harlen
David Fulton Publishers, 3rd ed. 2000

It's not your father's science class! The practice of inquiry-based learning—its nature, its importance, and how teachers go about it.

RECOMMENDED (JUST FOR JOY)

Uncle Tungsten
Memories of a Chemical Boyhood
Oliver W. Sacks
Alfred A. Knopf, 2001
Sacks's firsthand experience as a boy whose childhood was steeped in guided discovery. Not every mother's a scientist, as his was, but every parent can fan the fire of curiosity!

"What Do *You* Care What Other People Think?"
Further Adventures of a Curious Character
Richard P. Feynman
W.W. Norton, 2001

Feynman, in this memoir and its precursor *"Surely You're Joking, Mr. Feynman!"* takes us by the hand through a life of scientific and human surprises, disasters, joys, profundities, and discoveries.

SCIENCE AND MATH AT HOME

"At Home with Science" series
Janice Lobb

Larousse Kingfisher Chambers, 2000–2002

This rhyming-title series (*Dig and Sow: How Do Plants Grow? Splish Splosh: Why Do We Wash?*) brims with illustrations, riddles, science mysteries, and hands-on experiments for children aged 4–8.

Bouncing Eggs
Amazing Science Activities You Can Do at Home
McGraw-Hill, 1999

Using everyday materials, "Wizard Wellnitz" makes sleuths of 7- to 10-year-olds with exciting hands-on science activities and experiments.

FAMILY MATH
For Grades K–8
Jean Stenmark, Virginia Thompson, and Ruth Cossey
EQUALS, 1986

FAMILY MATH for Young Children: *Comparing*
For Grades PreK–3
Grace Dávila Coates and Jean Kerr Stenmark
EQUALS, 1997

FAMILY MATH—*The Middle School Years, Algebraic Reasoning and Number Sense*
For Grades 5–8
Virginia Thompson and Karen Mayfield-Ingram
EQUALS, 1998

These EQUALS classics are filled with engaging mathematics games, activities, and investigations for children and their parents to explore at home.

Home Science Kits from GEMS® and Scientific Explorer
LHS Discovery Corner
http://store.yahoo.com/lawrencehallofscience/sodpopscienk.html

Soda Pop Science…Jiggly Gems and Crystal Creations…Messages from Space…. Award-winning, stimulating family science activities that appeal equally to boys and girls, encouraging learning through exploration and invention.

"Science for Every Kid" series
Janice Van Cleave
John Wiley & Sons, 1993–2001

The books in this series (*Astronomy for Every Kid*, *Chemistry for Every Kid*, *Physics for Every Kid*, *The Human Body for Every Kid*, etc.) deliver the ideas, materials lists, and background science for hundreds of experiments children can perform safely at home.

The Brain Explorer
Puzzles, Riddles, Illusions, and Other Mental Adventures
Pat Murphy, Ellen Klages, Pearl Tesler, Linda Shore, and the Exploratorium
Henry Holt, 1999

San Francisco's renowned Exploratorium shares some of its most innovative science activities for families to do at home.

ACTIVITY WEB SITES

Lawrence Hall of Science Interactive Showcase
http://sv.berkeley.edu/showcase

Sparkling interactive math and science activities from some of the best minds in the business.

Math Activities for Parents and Their 2- to 5-Year-Old Children
www.ed.gov/pubs/EarlyMath/title.html

Simple activities you can do during the course of a day to stimulate your child to engage in math.

Nobel e-Museum
www.nobel.se

Scads of information about recent and past Nobel laureates, experiments and works of note, interactive games, and Nobel trivia.

LINKS FOR PARENTS

Education Departments by State
State Departments of Education and Alternative Schools
www.criminology.fsu.edu/jjclearinghouse.html

This very useful site was created, oddly enough, by the Juvenile Justice Department. It offers direct access, at the click of a link or a map, to educational services, departments, and institutions in nearly every state in the country.

Educational Resources Information Center (ERIC)
www.askeric.org

Administered by the National Library of Education (U.S. Department of Education), the AskERIC database offers over 3,000 resources on a variety of educational issues.

Families and Work Institute
www.familiesandwork.org

Advice, articles, and research on improving family life for working families. (Some products for sale.)

GreatSchools.net
www.greatschools.net

A marvelous resource for parents, this nonprofit organization provides valuable information about public, private and charter schools in all 50 states and detailed school profiles for Arizona, California, Texas, and Florida.

How to Have a Good Relationship with Your Child's Teacher
www.ldonline.com/ld_indepthparenting/hw_relationship.html

Useful advice to all parents on maintaining positive communication with their children's teachers.

Learning Network
What Works: Ideas from Parents
www.familyeducation.com/whatworks

Suggestions posted to this site range from "After School for Older Kids" to "Explore your Neighborhood Library." Worth an occasional look.

National Coalition for Parent Involvement in Education (NCPIE)
www.ncpie.org

NCPIE is devoted to the issue of keeping parents involved in schools as partners, and provides updates on federal legislation that affects this issue.

National Parent Information Network (NPIN)
http://npin.org

A project of the ERIC system, NPIN provides a question-answering service, toll-free reference numbers, addresses, and e-mail contact information for parents.

North Central Regional Educational Laboratory (NCREL)
www.ncrel.org

This nonprofit organization, through educational applications of technology, helps schools in the north-central U.S. build tools and apply proven practices so that students can reach their full potential.

Parent Portal
www.lhsparent.org

A wealth of hands-on suggestions, research digests, recommended resources, and feedback opportunities for adult caregivers of school-aged children.

INDEX

University of Durham, 76
University of Michigan, 76
Urban Parent Involvement, 35
U.S. Department of Education, 91, 92,
    99
USA Weekend, 78

Veblen, Thorstein, 97

*"What Do You Care What Other People
    Think?"*, 119
Will Standards Save Public Education?,
    59
World Wide Web, 87
    see also Internet

Yale Medical School, 17
YMCA/YWCA, 87, 103

About the Authors

Jacqueline Barber is Associate Director of the Lawrence Hall of Science (LHS), the world-renowned public science education center of the University of California at Berkeley. She has over twenty years' experience in education reform work and currently leads the LHS Center for Curriculum Innovation. She is author of many science and mathematics curriculum units and of *The Architecture of Reform* and *Parent Partners: Workshops to Foster School/Home/Family Partnerships*. She and her husband Steve live in Berkeley, California with their three sons, Saul, Leib, and Jesse, ages 12, 9, and 5.

Nicole Parizeau is a senior writer at the Lawrence Hall of Science. She is involved in science and mathematics curriculum development and led the creation of an online parent education project (www.lhsparent.org). She previously served as senior editor at *Whole Earth* magazine, has written for several publications in Canada and the United States, and was an interpretive naturalist for many years. She lives with her husband Jeff in Berkeley, California.

Lincoln Bergman leads a science and mathematics curriculum program at the Lawrence Hall of Science. He has coauthored numerous science and mathematics curriculum units and handbooks, and has written many poems that celebrate science and math. He was a radio news director and journalist in a past life. He and his wife Lisa live in Richmond, California, and he has two daughters: Anna, 23, singing and dancing in musical theater in New York City, and Caitlin, 15, facing the challenges of Berkeley High School. He volunteers in the high school's Writers' Room program, briefly described in this book.